Programmed Maintenance Capacity Assessment Tool

Workloads, Capacity, and Availability

Elvira N. Loredo, Raymond A. Pyles, Don Snyder

Prepared for the United States Air Force
Approved for public release; distribution unlimited

RAND PROJECT AIR FORCE

The research described in this report was sponsored by the United States Air Force under Contract F49642-01-C-0003 and FA7014-06-C-0001. Further information may be obtained from the Strategic Planning Division, Directorate of Plans, Hq USAF.

Library of Congress Cataloging-in-Publication Data

Loredo, Elvira N.
Programmed depot maintenance capacity assessment tool : workloads, capacity, and availability / Elvira N. Loredo, Raymond A. Pyles, Don Snyder.
p. cm.
Includes bibliographical references.
ISBN 978-0-8330-4015-2 (pbk. : alk. paper)
1. KC-135 (Tanker aircraft)—Maintenance and repair—Costs—Evaluation. I. Pyles, Raymond, 1941– II. Snyder, Don, 1962– III. Title.

UG1242.T36L68 2007
358.4'4—dc22

2006102497

Published 2007 by the RAND Corporation
1776 Main Street, P.O. Box 2138, Santa Monica, CA 90407-2138
1200 South Hayes Street, Arlington, VA 22202-5050
4570 Fifth Avenue, Suite 600, Pittsburgh, PA 15213-2665
RAND URL: http://www.rand.org/
To order RAND documents or to obtain additional information, contact
Distribution Services: Telephone: (310) 451-7002
Fax: (310) 451-6915; Email: order@rand.org

Preface

This monograph describes a model for evaluating the combined capacity of organic (U.S. Air Force–owned and –operated) and contractor maintenance assets to meet aircraft programmed depot maintenance (PDM) workloads. The PDM Capacity Assessment Tool (PDMCAT) forecasts the average number of aircraft that will be in PDM status each year over several decades,[1] based on the initial number of aircraft in PDM status, the physical capacity of the facility or facilities (number of docks available for conducting PDM work), the PDM induction policy (the period allowed between the completion of one PDM and the start of the next), and the minimum hands-on flow time (the minimum time it would take a facility to complete a PDM if only one aircraft were in PDM status). While not directly part of the model, the derived induction data can be used to estimate both near- and long-term obligation authority requirements for different induction policies, labor rates, and workload forecasts.

To illustrate the model's operations and capabilities, we applied the model to evaluate the U.S. Air Force's current capacity for supporting KC-135 PDM and examined several options for improving both near- and long-term availability. In the process, we discovered that, while future annual fleet costs increase and availability decreases with

[1] The Air Force tracks the operational condition and status of each aircraft from acquisition to disposal. When an aircraft is inducted into PDM (when the initial PDM tasks commence at an organic depot or contractor facility), it is in PDM status and is no longer available for training and operations until the PDM work has been completed and the aircraft has been transferred to the using command.

age and workload, they do so rather less rapidly because the aircraft induction rates (the number of aircraft inducted each year) decrease as the PDM flow time increases. This leads to a less-drastic cost and availability forecast than usual.

This monograph should be of interest to Air Force aircraft sustainment wings,[2] workload planners, PDM facility managers, cost analysts, long-term budget forecasters, and fleet replacement planners. It should also be of interest to analysts and modelers estimating the availability and cost effects of periodic maintenance activities, including systems ranging from commercial aircraft fleets to ships to vehicle fleets and even major building inspections and maintenance.

The work reported in this monograph was jointly sponsored by two projects within the Resource Management Program of RAND Project AIR FORCE. The PDMCAT model was developed in support of the Aging Aircraft Project, sponsored by Brig Gen David Gillett, then Director of Maintenance, Office of the Deputy Chief of Staff for Logistics, Installations, and Mission Support (AF/A4M), Headquarters United States Air Force. The application of the model to the KC-135 was sponsored by Brig Gen David J. Eichhorn, Aeronautical Systems Command Aircraft Enterprise Office (ASC/AA). This monograph continues work by Pyles (2003), which presents evidence of growth in maintenance workloads related to aging aircraft. The modeling techniques presented here are an extension one of the RAND coauthors, Don Snyder, made to the balanced job bound (BJB) model (Zahorjan et al., 1982). This extension of Zahorjan's work to include the multiple server case is presented in Appendix B. The technique presented here was also used in a KC-135 tanker recapitalization study (Kennedy et al., 2006).

---

[2] *Aircraft sustainment wing* is the new Air Force Materiel Command term for a system program director's office responsible for the engineering, material condition, airworthiness, and operational suitability of aircraft. We use that designation throughout this monograph.

A Note About the Data in This Monograph

Our study and an initial draft of this monograph had been substantially completed about the time that the KC-135 Analysis of Alternatives began. The publication of this monograph was postponed in deference to that more-comprehensive study. As a consequence, some of the data used in the analyses are now quite dated, and some scenarios discussed have been overtaken by events. Because our purpose is to describe the model and its potential application, these data and scenarios have been retained, even though the Air Force's plans for the KC-135 fleet have evolved substantially.

RAND Project AIR FORCE

RAND Project AIR FORCE (PAF), a division of the RAND Corporation, is the U.S. Air Force's federally funded research and development center for studies and analyses. PAF provides the Air Force with independent analyses of policy alternatives affecting the development, employment, combat readiness, and support of current and future aerospace forces. Research is conducted in four programs: Aerospace Force Development; Manpower, Personnel, and Training; Resource Management; and Strategy and Doctrine.

Additional information about PAF is available on our Web site at http://www.rand.org/paf.

Contents

Figures

Tables

Summary

Aging Air Force fleets have accrued material deterioration problems that have resulted in increasing maintenance workloads, which have, in turn, led to reduced availability of the fleets for operations and training. Nowhere has this problem been more apparent and severe than during the periodic inspection and repair of aircraft structural elements of PDM (see pp. 5–8).

PDM is conducted in large organic or contractor facilities where aircraft can be partially disassembled, inspected, and repaired. A typical PDM visit may require between 2,000 and 50,000 labor hours (depending on the fleet) and substantial material. The total labor required to complete PDM is expected to increase as a function of the age of the fleet. However, there are different perspectives on the form that this increase may take. One analytic community (which we refer to as *the engineers*) relies on engineering judgment and current planned workloads to theorize that future workloads might stabilize over the near term; another group (*the statisticians*) rely on statistically based cost and workload trends to theorize that workloads and costs will grow and that availability will decrease.

Traditional Modeling Approaches Have Limited Applicability

While detailed resource and process simulation models can be constructed for a specific facility at a specific point in time, the workload, processes, and resource availabilities change constantly. More problematic, the specific workflows used by competing entities (organic or contractor) are seen as a proprietary matter that affects their ability to compete for future workloads. As a consequence, few facilities are willing to share detailed information on their specific work processes.

We developed PDMCAT to be able to estimate the number of aircraft in PDM status, future inductions, and production levels and to rely only minimally on detailed information from inside a facility (see pp. 9–12). We also sought to rely on easily observable features, such as the number of docks for performing maintenance and recent measures of actual performance, so that having "inside" information was not critical to forecasts of future inductions or numbers of aircraft in PDM status (i.e., not available for operations and training).

To that end, we extended and elaborated the BJB model (Zahorjan et al., 1982) to include multiple servers within each job stage. The original model was developed for the operational design of computing time-sharing systems. Appendix A discusses queuing theory related to this model. The BJB model required very little information in the first place, and we were able to simplify its data requirements further and apply it to the PDM process. Chapter Three describes application of the model and its development; Appendix B presents more detail on our extension.

Testing and Demonstrating PDMCAT: The KC-135 Case

To test and demonstrate the model's capabilities, we applied it to the KC-135 PDM process described in Chapter Four, first examining how well the model was able to forecast recent PDM performance, then comparing two alternative forecasts of the future workload and evaluating capacity and PDM process-improvement options to maintain acceptable availability levels. That fleet was chosen because there was an ample amount of information about its recent workloads, number of aircraft in PDM status, and changing capacity. More important, that fleet had experienced a substantial change in the number of aircraft in PDM status during the years 1998–2002, so we believed it would constitute a good test of the PDMCAT model's forecasting capabilities (see pp. 13–22).

The alternative forecasts reflected engineers' versus statistical workload predictions. The fleet reduction program example demonstrates how changes in fleet size would reduce the number of aircraft in PDM status as the number of aircraft inducted each year diminishes. Figure S.1 shows the aircraft purpose possession history of the KC-135 tanker fleet from the second quarter of fiscal year 1995 to the first quarter of fiscal year 2004.[3] This chart shows the increase in the so-called *depot-possessed aircraft* and the consequential decrease in that

Figure S.1
Changes in Depot Capacity and Required Workload Created a Bubble in Depot-Possessed Aircraft

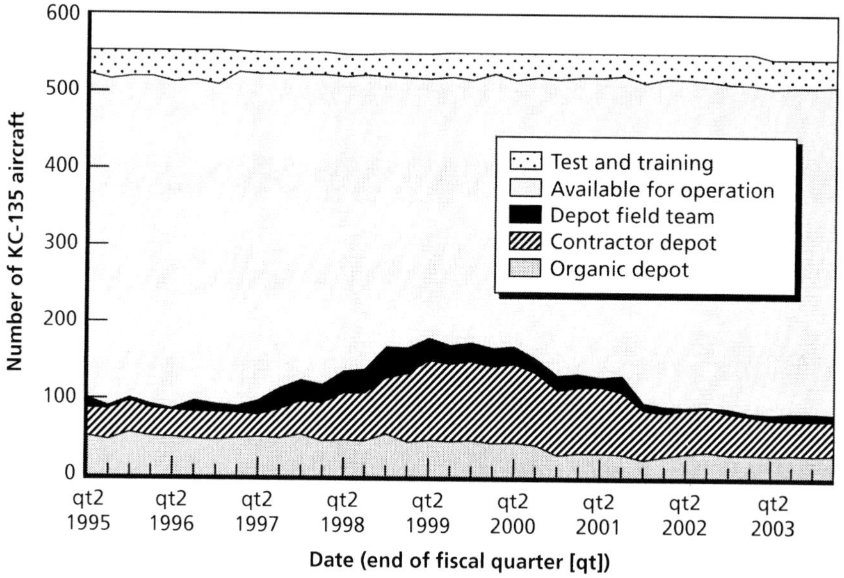

RAND *MG519-S.1*

---

[3] The aircraft purpose possession history indicates how many Air Force aircraft are possessed for different purposes (e.g., test, training, modification, maintenance). It is constructed from detailed daily possession status change reports for each aircraft serial number. Most important for this study, it contains information from which one can compute the historical number of aircraft in PDM status and the number that entered PDM each year.

aircraft's availability for operations starting in the third quarter of 1997 and peaking in the second quarter of 1999—with almost 200 KC-135 tankers either in possession of depot field teams or at organic or contractor depot facilities. Our initial analyses addressed the PDMCAT model's ability to replicate that experience.

Initial Analysis of the PDMCAT Model

We used historical workload data to compare the model's forecasts to actual aircraft in PDM status during a critical transition period—from 1997 through 2003. During this time, the number of aircraft in PDM status increased by more than 50 percent, then returned to levels below the initial 1997 level. Figure S.2 shows that the PDMCAT model accurately reflected the increase and subsequent decrease in aircraft in PDM status.

Figure S.2
PDMCAT Forecasts Using Actual Workloads Match Actual In-Work Forecasts Using the PDMCAT Model

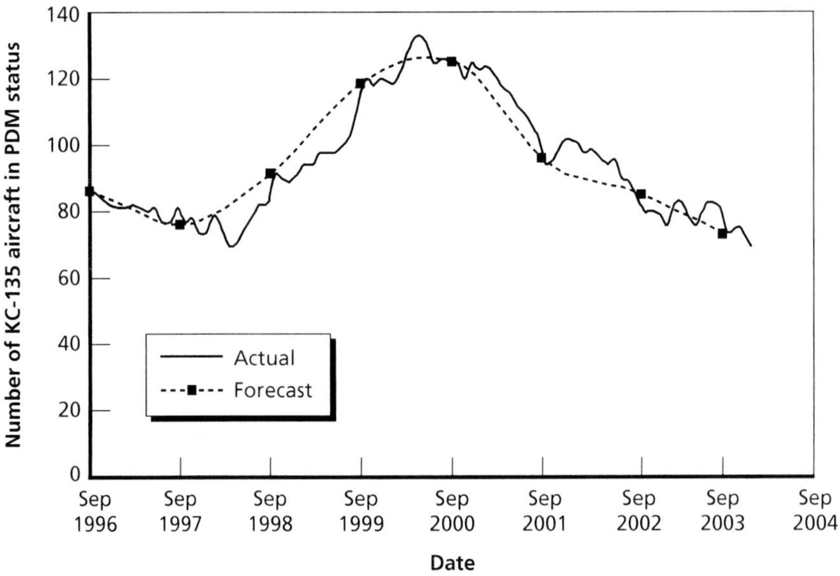

RAND *MG519-S.2*

Using the PDMCAT Model to Assess Assumptions About Future Operations

Finding the historical match acceptable, we applied the model to test how assumptions about workload plans, induction schedules, labor application rates (often called *burn rates*), depot capacity, and fleet size would affect the forecast of near- and long-term inductions, production quantities, and aircraft in PDM status. A sample of how we used PDMCAT to test various assumptions is shown below.

Forecast of Future Workloads

Two forecasts of future PDM workloads were used in the Chapter Four analyses. The first, developed by the *KC-135 Economic Service Life Study* (ESLS) (Sperry et al., 2001), uses both statistical analysis and expert engineering judgment to estimate the effect of fatigue cracking and corrosion growth on future PDM workloads. The second is a purely statistical equation drawn from a PAF study that sought to discover and characterize maintenance life-cycle workload patterns that were common across all Air Force fleets, rather than a pattern that may reflect some idiosyncratic temporary behavior in a single fleet's history (Pyles, 2003). (See pp. 18–21.)

We used the model to examine both near-term (one to five years) and long-term PDM performance. In the near-term cases, we assumed there was only limited opportunity to increase PDM capacity, but that the PDM induction policy (i.e., the interval between subsequent PDMs) could be used to manage the workflow and aircraft availability. In the long-term cases, we assumed that it would be possible to add physical capacity (docks where aircraft could receive PDM maintenance) and to introduce process improvements that could increase the labor application rate (the number of labor hours that can be usefully applied to a single aircraft in a single day). (See pp. 21–24.)

Using PDMCAT to Moderate the Effects of Changes in Aircraft Induction Intervals on Near-Term Work in Process

The KC-135 fleet PDM process has experienced a turbulent period during which previously stable flow times and production rates were disrupted by a period of low production outputs followed by a period of higher-than-usual production outputs. If the KC-135 PDM managers were to follow Air Force Technical Order (AFTO) 00-25-4 (U.S. Air Force, 2003) interval prescriptions exactly, those production fluctuations would reappear as induction fluctuations, creating an imbalance between depot capacity and incoming workload requirements (see pp. 26–29). PDM managers have some leeway in adjusting aircraft induction intervals. This was the case in 2002 and 2003, when we found that the depot inducted five more (in 2002) and 28 more (in 2003) aircraft than required by AFTO 00-25-4 (see U.S. Air Force, 2003).

Figure S.3 shows how we used the PDMCAT model, along with the PAF workload forecast, to demonstrate the effect of those early inductions on aircraft in PDM status in subsequent years. Over the near term, the model projects a temporary reduction in the number of aircraft in PDM status, followed by an equally temporary increase in that number that would begin to approach the peak number of aircraft in PDM status from 1997 through 2003. The later increase was caused by a forecast increase in PDM workload coinciding with the scheduled return to PDM of the additional aircraft produced in 2003–2004. By adjusting the annual induction rates during these periods, we were able to use the model to identify an alternative induction plan that would reduce the peak number of aircraft in PDM status to acceptable levels through 2009.

Using PDMCAT to Test Assumptions About Long-Term Workload Growth, Increases in Capacity, and Burn Rates

Looking to the long term, which is depicted in Figure S.4, we found that the more pessimistic PAF workload projection would cause the depot flow times to increase until the "aircraft in PDM" status would

Figure S.3
PDMCAT Near-Term Forecasts Modulated by Changing Inductions

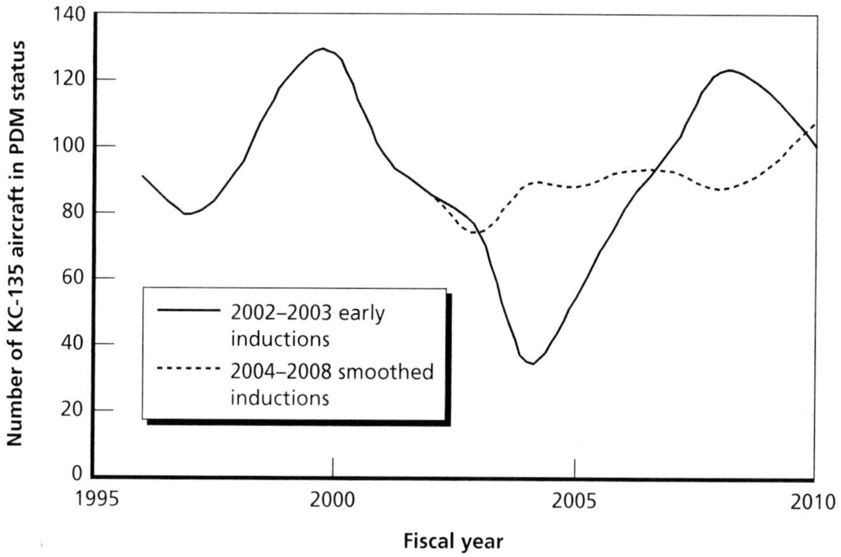

RAND *MG519-S.3*

reach the 1997 to 2003 peak by 2013. We then increased either the physical capacity (number of docks where maintenance can be performed) or the labor application rate (a composite factor reflecting both labor available across all shifts and the degree of parallel operations in the PDM process) by 50 percent in 2010 to evaluate how those capacity increases might change the availability forecast. We learned that the increases both reduced the number of aircraft in PDM status and prolonged the time until the 1997 to 2003 surge peak was reached. The labor application rate option performed better, not reaching the 1997 to 2003 peak until 2024, compared to 2020 for the capacity increase case. We next examined the implications of the ESLS engineering-based workload forecast, which yielded a much more optimistic long-term outcome, never quite reaching the 1997 to 2003 peak (see pp. 32–43).

Figure S.4
Adding Capacity and Increasing the Labor Burn Rate Delay Impact of PAF Workload Forecast

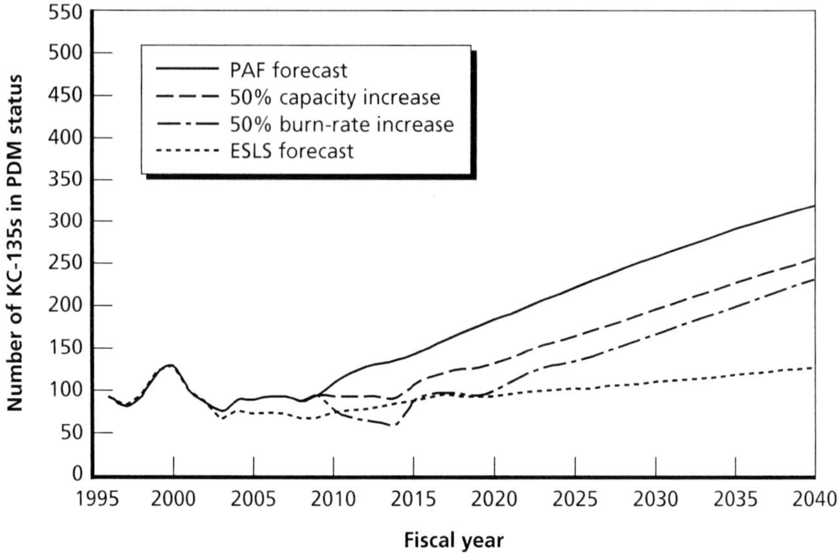

RAND *MG519-S.4*

Using PDMCAT to Forecast the Effect of Changes in Fleet Size

In Chapter Five, we compared the PAF forecast against the KC-135 system program director's (SPD's) engineering-based forecast (see p. 20), assuming that the Air Mobility Command (AMC) plan to retire KC-135Es would have been implemented until only 490 aircraft remain in the fleet: 417 KC-135R/Ts and 73 KC-135Es.[4] We further assumed that the capacity would change in proportion to changes in the projected workloads (see pp. 47–49). With the KC-135 Tanker Sustainment Group's moderate forecast of PDM workloads, the PDMCAT

[4] This plan was not implemented, but the analysis sheds light on how it would have affected KC-135 aircraft availability.

model projects that the aircraft in PDM will not reach the 100-aircraft level until after 2050.[5] Under the less optimistic PAF forecast, the PDMCAT model projects that the number of aircraft in PDM status will reach 100 as early as 2013, even if the fleet size is reduced as planned. This projection is contrasted with the results shown in Figure S.5 (PAF forecast 1). The conjunction of reducing the KC-135 inventory and increasing capacity significantly reduces the effect of increased workloads on aircraft availability.

Figure S.5
Reducing KC-135 Inventory and Increasing Capacity Dampen Surge in Aircraft in Work

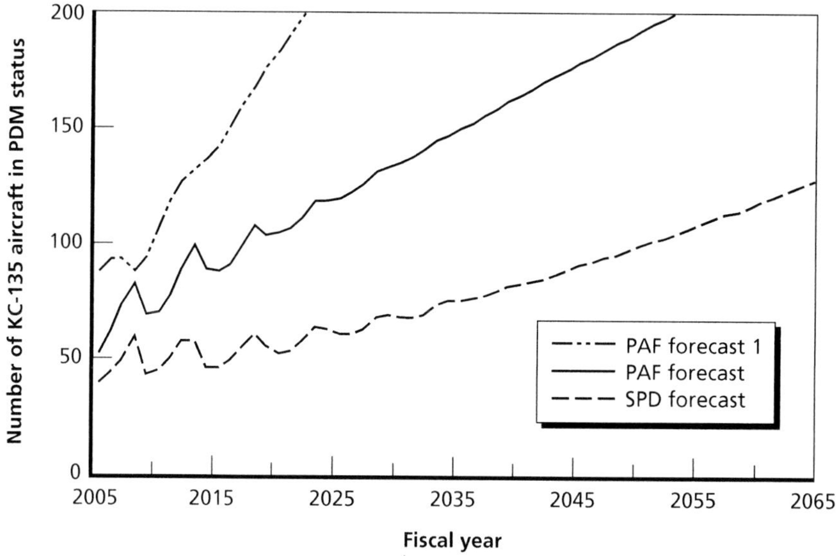

RAND *MG519-S.5*

---

[5] The office's formal designation has recently been changed from the KC-135 SPD office to the 437th Tanker Sustainment Group (437 TSG). The forecast was very similar to that for the KC-135 ESLS but was based on more-recent decisions that eliminated some near-term tasks and postponed others.

Limitations of the PDMCAT Model

PDMCAT is a macro-level forecasting model. As with all forecasting models, it is sensitive to the accuracy of the factors used to generate the forecast. PDMCAT requires three critical factors: a forecast of future workloads, an estimate of the maximum labor application rate, and an estimate of the depot capacity.

Future PDM workloads are the subject of some debate. Pyles (2003) found a general cross-fleet pattern for PDM growth as fleets age and a significant second-order term related to age. An analysis focused solely on the KC-135—the KC-135 ESLS (Sperry et al., 2001)—also projected continued growth on KC-135 PDM workloads, although at a less pronounced rate than that found by Pyles. The 437 TSG workload forecast closely mirrors the ESLS forecast in terms of rate of growth, but projects fewer hours per PDM. The PDMCAT forecast of aircraft in work will vary depending on the workload forecast used. While workloads have grown in recent years, this is hardly conclusive evidence that the trend will continue into the future. Some argue that the workload growth will necessarily taper off as all or most of the key components on the KC-135 are repaired or replaced. Therefore, users of the PDMCAT model to forecast long-term trends in aircraft availability (20 or 30 years into the future) should periodically review and refine the available workload forecasts to reflect more-recent information that may reduce those differences in workload forecasts (see pp. 46–49).

An estimate of the maximum labor application rate (sometimes called the maximum hands-on burn rate), the rate at which labor can be applied to PDM workload, may change over time as process improvements, learning, and technology allow depot personnel to work more efficiently. As it becomes possible to apply more labor simultaneously to each aircraft, the PDM flow times will diminish. However, some changes in the underlying processes, such as subcontracting some tasks to outside entities, may reduce both the measured workload and the measured maximum labor application rate without necessarily reducing the flow time as the PDM process waits for the completion of subcontracted work. When workloads are contracted out or otherwise moved from the formal PDM package, it is important to reestimate the labor application rate. As an estimate of depot capacity, the PDMCAT model measures depot capacity in docks—the number of aircraft that can receive work simultaneously at the maximum labor rate. The modeler has the option of entering a constant number of docks or of increasing the number of available docks over time. However, the PDMCAT model does not assess how the addition of docks may change the labor skill mix and affect the burn rate, nor does it consider how additional docks are added. That is, PDMCAT does not differentiate the addition of docks within an existing facility (by freeing up space currently occupied by other workloads) from the addition of docks by hiring contractors or by otherwise increasing physical capacity.

Although the underlying mathematics of the PDMCAT model support both lower- and upper-bound calculations on PDM throughput, the model produces only an estimate of the upper bound. Estimating the lower bound requires additional information about the imbalances across various stages of the PDM processes (i.e., the times and resources devoted to different PDM tasks) that would seldom be available to an external observer because of the competitive value of that information. In addition, we assume that PDM process managers will allocate their resources across those tasks to maintain a balanced production process, in which the average throughput rates at each production stage are roughly equal.

Conclusions

We were able to use the model to examine some important near- and long-term issues associated with the KC-135 fleet. While we were impressed with the model's existing capabilities, we have already begun to extend it to deal with multiple fleets using shared facilities, fleets with induction periods of less than a year, and modification workloads (see pp. 56–57).

With regard to the KC-135, we found that the shapes of the availability and cost forecasts did not grow in proportion to workloads, as assumed in many studies.[6] Future studies forecasting PDM costs and aircraft availability may need to consider using PDMCAT or equivalent calculations to estimate how changing PDM workloads will affect fleets' budgets and availability (see pp. 53–55).

---

6 The KC-135 Analysis of Alternatives study (Kennedy et al., 2006) is an exception. A version of PDMCAT was used to estimate the number of aircraft in PDM and modification status, and the PDM costs associated with several different workload forecasts.

Acknowledgments

This work has benefited from feedback and comments from several of our RAND colleagues, including Edward Keating, C. Robert Roll, Jr., and Timothy Ramey. We also appreciate the support and encouragement of Karl Hart, of Alion Corporation, supporting ASC/AA.

The authors would also like to thank the reviewers, Gregory Hildebrandt and Richard Hillestad, for their insightful and helpful critiques.

Abbreviations

ABA	asymptotic bound analysis
AETC	Air Education and Training Center
AFKS	Air Force Knowledge System
AFMC	Air Force Materiel Command
AFRC	Air Force Reserve Command
AFTO	Air Force Technical Order
AMC	Air Mobility Command
ANG	Air National Guard
ASC/AA	Aeronautical Systems Command Aircraft Enterprise Office
BASC	Boeing Aerospace Support Center
BJB	balanced job bound analysis
BRAC	Base Realignment and Closure
CIE	controlled interval extension
CTMC	continuous time Markov chain
DoD	U.S. Department of Defense
DPEH	depot product earned hours

DPSH	depot product standard hours
DynaSIM	Dynamic Simulation of Intermediate Repair
ESLS	*KC-135 Economic Service Life Study*
FY	fiscal year
LCOM	Logistics Composite Model
MAJCOM	major command
MSR	major structural repair
PACAF	Pacific Air Forces
PAF	Project AIR FORCE
PDM	programmed depot maintenance
PDMCAT	Programmed Depot Maintenance Capacity and Assessment Tool
PERT	Program Evaluation and Review Technique
PPBS	Planning, Programming, and Budgeting System
SM-ALC	Sacramento Air Logistics Center
SPD	system program director
TSAR	Theater Simulation of Airbase Resources
TSG	tanker sustainment group
UDLM	unscheduled depot-level maintenance
USAFE	U.S. Air Forces in Europe

CHAPTER ONE

Introduction

One of the greatest concerns arising from current U.S. Air Force aircraft replacement plans is the potential deterioration of aircraft availability as maintenance workloads grow due to age-related material deterioration. Nowhere is that growth more troubling than in the workloads associated with programmed depot maintenance (PDM). In at least one case—the KC-135 tanker fleet—the growth in PDM workloads recently exceeded the combined capacity of the organic and contractor organizations that conduct PDM activities and caused an unacceptable reduction in the number of aircraft available for operations and training.

That situation arose without warning. The Air Force does not currently have a method that can help predict future imbalances between workload and capacity and cannot evaluate the implications of such an imbalance for aircraft availability. Not being able to assess how changes in depot capacity and PDM workloads may impact aircraft availability increases the potential that the Air Force may not have adequate numbers of aircraft available for operations and training.

Changing Demands of PDM Assessments

Previous attempts to assess PDM capacity have been stymied by the sheer complexity and inaccessibility of data concerning the PDM process. PDM activities for a typical large aircraft may include thousands of technicians with dozens of skills spread across one to three different main facilities served by dozens of local and remote subcontractors and material providers. The workload content varies substantially across different aircraft and evolves as new material-deterioration modes emerge. Even more challenging, the maintenance process, equipment, facilities, skills, subcontractors, and material providers also fluctuate constantly to respond to those changing demands.

The Air Force needs a way to appraise its aggregate PDM capacity that does not require the voluminous, up-to-date information required by the traditional approaches. This document describes such an approach, using a simple model to relate depot maintenance workloads and capacity to aircraft availability.

Organization of This Monograph

This monograph is organized as follows. Chapter Two describes the depot process and introduces the modeling method. Chapter Three presents a summary of the KC-135's PDM history and our approach to estimating the model parameters. Chapter Four illustrates the use of the Programmed Depot Maintenance and Capacity Assessment Tool (PDMCAT) to compare alternative PDM workload and capacity scenarios and their effects on availability and costs. Chapter Five examines the Air Mobility Command (AMC) plan to reduce the existing fleet size by retiring KC-135Es until the fleet size reaches 490. Chapter Six presents conclusions and next steps. Appendix A presents an explanation of the relevant queuing models and theory. Appendix B presents RAND's expansion of Zhorjan's Balanced Job Bound (BJB) analysis to the multilevel case. Appendix C presents a detailed discussion of how the PDMCAT parameters are estimated, and it includes a discussion of how PDM induction policies and fleet retirements might impact model results.[1]

[1] Air Force Technical Order (AFTO) 00-25-4 (U.S. Air Force, 2003) specifies fleet-unique maximum intervals between PDMs. Those intervals measure the time allowed between the completion of one PDM and the commencement of the next. For example, most KC-135 aircraft must be inducted into (i.e., enter) the PDM process within 60 months of completing the last PDM. However, aircraft operating in the Pacific theater must enter every 48 months. There is some flexibility in the interval, in that an aircraft's interval can be extended by six

A Note About the Data in This Monograph

Our study and an initial draft of this monograph had been substantially completed about the time that the KC-135 Analysis of Alternatives (AoA) began (Kennedy et al., 2006). The publication of this monograph was postponed in deference to that important study. As a consequence, some of the data used in the analyses are now quite dated, and some scenarios discussed have been overtaken by events. Specifically, the workload, induction, and availability data used for comparisons in Chapters Three and Four end in 2003, and the fiscal year (FY) 2004 KC-135E retirement plans evaluated in Chapter Five were never executed. Because our purpose is to describe the model and its potential application, these data and scenarios have been retained, even though the Air Force's plans for the KC-135 fleet have evolved substantially.

---

months, provided that a field inspection has been conducted to verify that the aircraft can be operated safely for that extended period.

CHAPTER TWO

Background and Theory

As discussed in Chapter One, the PDM process is complex and constantly evolving. This combination of problem characteristics makes analyzing PDM performance particularly challenging. In this chapter, we describe the PDM process and how it affects the availability of aircraft for operations and training.

The PDM Process

The primary purpose of Air Force–owned military aircraft is to support operations and training. However, those activities place stresses on the aircraft structure and systems whose material-deterioration effects are cumulative in nature, especially metal fatigue and corrosion. Thus, it is necessary for the Air Force to set aside some regularly scheduled interval during which each aircraft is substantially dismantled, inspected for unsafe material, and restored to safe operating condition. Depending on the size of the aircraft, its design, its material, its age, and the operating stresses it encounters, an Air Force aircraft will spend from three to 18 months in PDM every three to eight years. To manage that process, the Air Force formally transfers possession (custody) of each aircraft from the major commands (MAJCOMs) to an organic (Air Force Materiel Command [AFMC]–owned and –operated) depot facility or a contractor facility for a specified list of maintenance tasks that make up "basic" PDM.

In the course of that basic PDM, the organic or contractor facility may discover material deterioration that requires additional effort to remediate, and that work is generally performed as soon as authorization is given and funding is approved. Once all the approved maintenance tasks are completed and the reassembled aircraft is tested for airworthiness and functional operability, the Air Force formally transfers possession back to the owning MAJCOM.

The interaction between aircraft possessed by the MAJCOMs and those possessed by the depot is shown in Figure 2.1.[1] To make the description concrete and interesting, we have taken a specific example based on the KC-135 depot maintenance process at the Oklahoma City Air Logistics Center (Johnson, 2000). While the depot maintenance process for other aircraft may differ in the capacity of the facilities, procedural details, and processing times required, they are procedurally similar. In addition to basic PDM work, the depots also perform unscheduled depot-level maintenance (UDLM) and modifications. UDLM jobs are referred to the depot because they are beyond the technical capability of base maintenance. This type of work is unscheduled and unpredictable. For example, lightning strikes or rough landings are common causes of UDLM. Modifications are intended to improve capability or safety or to meet new regulatory requirements. Some modifications are performed concurrently with PDM, while others are scheduled separately and performed by depot field teams or at contractor locations. The modification workload is an important component in predicting aircraft availability.[2]

Figure 2.1 shows that a fleet's aircraft cycle between individual MAJCOMs and the depot. The period between depot visits is set by engineers and is based on the fleet's design characteristics and utilization, its previous maintenance history, and evaluations of how long a typical aircraft from the fleet can operate without undergoing inspection for flight-safety–critical material deterioration. Because utilization and

---

[1] Throughout this monograph, we use the generic designation of *depot* to encompass both the organic and contractor shops that perform PDM work.

[2] In general, modification work is expected to have a slower labor application rate than work performed under the basic PDM package. The *labor application rate*, sometimes called the *burn rate*, is a measure of how many productive labor hours the facility can generate in a single day on a single aircraft.

Figure 2.1
Depot-Level Work Flow

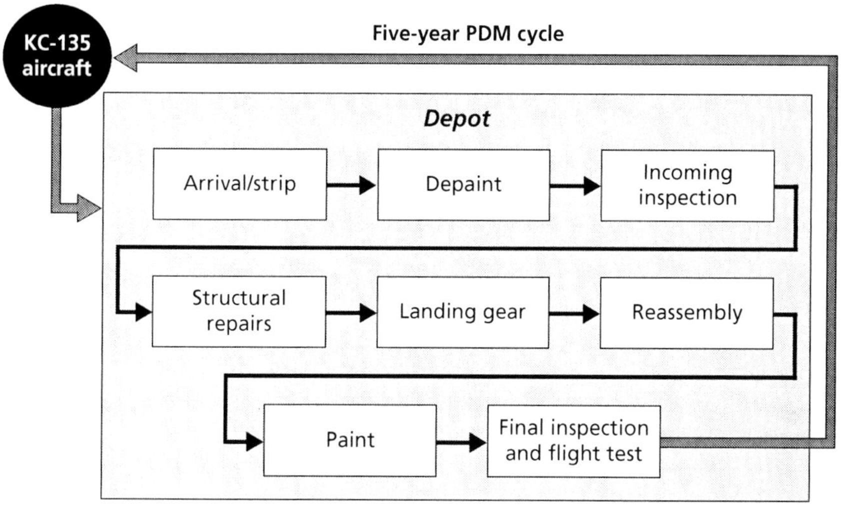

RAND *MG519-2.1*

maintenance effects are cumulative, and because engineers' knowledge and confidence in flight safety characteristics change over time, that period is adjusted as the fleet ages. In the past, the inter-PDM period for most fleets increased as they aged.

The PDM process shown in Figure 2.1 is a closed-loop network of stages through which each aircraft will pass in series.[3] The aircraft are possessed and operated by the MAJCOMs until they reach the end of their scheduled PDM period. After entering PDM, each aircraft proceeds through sequential stages, including arrival/strip, depaint, incoming inspection, structural repairs, landing gear, reassembly, paint, and final inspection and flight test. Each stage has one or more

---

3 *Closed-loop* (or *cyclic*) networks have special structures that differentiate them from open networks. Specifically, in *closed* networks, finite number of jobs (aircraft, in this case) cycle through the network. In an *open* network, jobs may enter the network randomly from the outside at specified entry points and depart the network from various exit points. This is not the case in a closed network. These properties led to the development of specialized queuing models that account for the conservation of jobs flowing around the network. That is, the number of jobs (in our case, aircraft) in the system is specified and changes slowly, if at all. For details, see Bolch et al. (1998).

docks at which aircraft are parked until they are ready to move on to the next stage. The number of docks at each stage varies according to the required workload. Once an aircraft is parked at a dock, it does not move to another dock within that same stage, and all aircraft movements occur only from one stage to another.

The first stage in the KC-135 PDM process shown in Figure 2.1 is Arrival/strip. Here, the PDM facility prepares the aircraft for the detailed inspections by defueling the aircraft and removing interior equipment and material, major components, and subsystems, either for their own inspections and repair (e.g., landing gear) or for storage elsewhere (e.g., engines, flight controls). In the case of the KC-135 aircraft, for the PDM flow depicted in the figure, the aircraft also has its paint removed to facilitate the exterior inspection of its structural components.

Next, inspection teams open and examine all areas of the aircraft according to a work specification that is updated annually to reflect predicted and observed material-deterioration patterns.[4] When minor repairs are required to control or remedy some deterioration, the inspection team and skilled specialists often perform the repair immediately. The results of the inspection inform the work conducted during the PDM and are used to periodically update future work plans.

More-extensive repairs (e.g., replacing a major structural element such as a skin panel) are undertaken in the structural repair stage, in which aircraft can remain for relatively long periods (up to several weeks) without interfering with work on other aircraft. The repair of major structural elements limits the aircraft mobility and other maintenance activities undertaken during that period. Once the major structural repairs are complete, the aircraft's landing gear and other equipment are reinstalled, and the aircraft is painted. Finally, the aircraft is

---

[4] Some deterioration patterns can be predicted, but only approximately. In particular, elaborate computer models with adequate information about aircraft usage can predict the growth of a fatigue crack, given its initial flaw size. By assuming a conservative (i.e., large) initial flaw, engineers can schedule inspections in anticipation of the flaw reaching a critical size at which flight safety is jeopardized. Models that can predict the effects of corrosion on structural integrity are not yet generally accepted by the engineering community.

subjected to a series of static and flight tests before being returned to the using command.

In recent years, concern has been raised that the PDM workloads and material requirements may grow as fleets age, leading to increased operating costs and declining availability.[5] Analyses of historical data (Pyles, 2003) and detailed engineering assessments of individual aircraft fleets (Sperry et al., 2001) have confirmed that likelihood.

Modeling the PDM Process

Most models that have been used to forecast aircraft availability have estimated the number of aircraft undergoing maintenance (or awaiting parts),[6] then subtracted that number from the total inventory to compute the number (or percentage) of aircraft available for operations. We have also adopted that technique, so we estimate the number of aircraft in the PDM activity and subtract the result from the total aircraft inventory.

Our approach differs from others in its application of a queuing theory model known as balanced job bound (BJB) analysis. Traditional applications of queuing theory require access to detailed information about the existing process. An example of this type of model is a sortie-generation model presented by Dietz and Jenkins (1997). This model incorporates multiserver queues and routing probabilities that require a detailed knowledge of the times and probabilities of different activities involved in aircraft flow through a network. In their particular case, the activities include taxiing (positioning the aircraft for takeoff), sortie (actually flying the aircraft for some period of time), turnaround

---

[5] A U.S. General Accounting Office (now Government Accountability Office) report (GAO, 2004) identified five primary factors that showed why the Air Force depot maintenance activity group's average price increased from $119.99 per direct labor hour of work in FY 2000 to $237.84 per hour in FY 2004. An increase in material costs accounted for about 67 percent of the total increase and was by far the most significant factor. The Air Force has identified some of the causes of the higher material costs, such as aging aircraft, but has yet to complete an effective and comprehensive analysis of material cost increases.

[6] In this initial work, we set aside the issue of aircraft not available for parts.

(refueling and scheduled maintenance), munitions upload, and numerous alternative repair activities (unscheduled maintenance).

Our implementation of BJB uses the concepts discussed by John Zahorjan et al. (1982) and modified by RAND to incorporate the multiple server case. See Appendix A for a discussion of BJB and Appendix B for details of RAND's modification. Zahorjan's work was developed for use in the design of early computer systems that contained a number of unique resources (specific disks containing specific files, tape transports, dedicated processors for specific tasks, transmission channels, and remote users with keyboards and displays). The principle was to develop a model that could be applied to test a proposed network's design without having to know many of the details, such as transition probabilities, which are often required by other queuing models.

Zahorjan and his colleagues developed upper and lower bounds on the number of jobs that a computer network could process (throughput), assuming a single server at each node of the network. Zahorjan's analysis depends on a critical mathematical observation: Any closed system's throughput is bounded by the performance of two other, theoretical, "balanced" systems. By *balanced*, we mean a system in which each job entering the system spends the same amount of time (Zahorjan calls this the server's *load*) at each server in the system. The upper production bound depends on the average load across all servers, and the lower bound depends on the maximum load. As these two values approach each other, the bounds become increasingly tighter.

Of course, this effect has always been recognized by production and maintenance shop designers and managers who strive to eliminate bottlenecks by adding or rebalancing resources and redesigning job steps. As we describe later in this monograph, we assume that PDM shop managers strive to minimize the chance of bottlenecks by balancing workloads across the shop's docks. This assumption enables us to use Zahorjan's upper bound as a production-estimation tool, even though we cannot imagine how Figure 2.1's detailed stage-by-stage workloads, flow times, and docks may ebb and flow over time.

But Zahorjan's observation implicitly placed a severe mathematical restriction on the applicability of his model: Every job had to require the same average amount of time on each server on every cycle around the system. That is, there could not be two or more interchangeable servers providing the same service, only one of which provides the service on any given cycle. In the PDM case shown in Figure 2.1, every production stage has multiple, interchangeable docks that can provide the same service.

RAND expanded Zahorjan's BJB model to include multiple servers within each job stage. Like Zahorjan, RAND derived both upper and lower mathematical bounds on throughput. Unfortunately, RAND's lower bound includes a term that places a high penalty on having a large, complex production system with many servers per stage. Essentially, the lower bound includes a fixed delay unrelated to queuing or actual work time that depends only on the number of servers per stage.[7]

More important in the PDM context, RAND's lower bound requires information that is not available during long-term planning—the load on the most severely loaded stage.[8] While there are aggregate models to predict total workload in the facility with some confidence over longer periods, no such models exist for the work within each stage. More important, there are no models to predict how shop managers will respond to those changes by reallocating workload or docks to rebalance the load at each stage. Thus, one cannot compute the maximum load.

Fortunately, we have observed that managers try to balance workloads across the system as much as possible. While a perfect balance may be difficult or impossible to achieve due to the technical process requirements at each stage, we anticipate that these continuous management adjustments will approach the performance estimated by the Zahorjan-RAND upper bound.

Putting the Zahorjan-RAND model into the context of a PDM facility, we get nodes that represent stages and servers that represent

---

[7] Essentially, the lower bound makes the conservative assumption that the aircraft must wait at each stage until all the docks are idle before it can receive service. While the assumption makes the mathematics tractable, it is not an accurate reflection of the actual process.

[8] As described by RAND, the load is the time required to complete each stage, divided by the number of servers in that stage. The maximum load in the system would then be the maximum load across all stages.

docks where aircraft are parked and work is performed. For example, in Figure 2.1, "Paint" is a major work center; it may have two docks (servers) at which aircraft are painted in parallel. In a depot, some mechanics may be assigned to stages according to skill, while others may move from stage to stage, or from dock to dock, depending on a manager's desire to expedite work and balance workloads.

Chapter Three discusses how we use available PDM process data to derive the information needed and how a model we call *PDMCAT* uses this information to set bounds on aircraft in work in PDM.

CHAPTER THREE

Using the Model: Obtaining Relevant Data and Designing Cases for Assessment

PDMCAT users face two complementary challenges: obtaining data to assess how the available PDM capacity will affect a fleet's future availability and designing cases that provide insight into ways to change those outcomes. This chapter addresses those challenges in turn by applying the model to the KC-135 fleet, so we describe the recent history of KC-135 PDMs to provide a background for that application.

KC-135 PDMs Have Undergone Recent Changes

The period from 1995 through 2003 was one of great turbulence in the KC-135 production process. In September 1995, Congress accepted the Base Realignment and Closure (BRAC) Commission's report that recommended closing the Sacramento Air Logistics Center (SM-ALC), among other U.S. Department of Defense (DoD) facilities. To arrange for the transfer of all the SM-ALC workloads to other facilities, the Air Force held a competition in which all the SM-ALC workloads were bundled, or combined, into a single contract up for bid. Thus, the KC-135 PDM workload from both SM-ALC and PEMCO (an overhaul and maintenance contractor company based in Birmingham, Alabama) was combined with all the other work previously performed at SM-ALC (a mixture of component repair; ground radar system support; and support of older, "mature" weapon systems, such as the A-10 and F-111), and AFMC solicited bids for the combined workload. The

Boeing Aerospace Support Center (BASC) facility and Ogden Air Logistics Center (OO-ALC) teamed in a novel public-private partnership and won the bundled contract in 1998. The two partners then directed non–KC-135 workload to OO-ALC and the KC-135 PDM and modification work to the BASC facility in San Antonio, Texas (Bergren, 2001). KC-135 work commenced at the San Antonio facility in October 1998. Throughout this period, KC-135 PDM work continued at the Oklahoma City Air Logistics Center (OK-ALC).

PEMCO was also undergoing a turbulent period (Chandler, 2001). Lacking a partner for the non-PDM workloads, it was unable to bid on the bundled workload. In an attempt to make its operation more competitive, the company requested work-rule changes from its union that triggered an eight-month strike beginning in July 1996. In October 1999, the company changed hands, new labor contracts were negotiated, and new financing was obtained. As the previous KC-135 PDM contract with PEMCO was about to come to an end and BASC had not yet attained full production capacity, the Air Force granted PEMCO a 36-aircraft extension through 2001 to supplement the initially low production rate achieved at the BASC facility. During that time, PEMCO negotiated a subcontract with BASC to perform a portion of the KC-135 PDM work.

All this turbulence took its toll on the total KC-135 productive capacity. Production outputs fell, but aircraft that required PDMs continued to arrive according to intervals specified in AFTO 00-25-4. Thus, the combined facilities' work in process and depot field team maintenance grew steadily from 1996 until 1999, as shown in Figure 3.1.[1] In FY 1997 and FY 1998, the work in process at the PEMCO facility doubled. By mid-1999, the organic and contractor

---

[1] By the third quarter of FY 1996, over a dozen aircraft were in field maintenance; by the fourth quarter of FY 1998, the figure rose to over three dozen. The underlying causes of those field maintenance activities are unclear, but many of the individual events occurred within six months of PDM on the same aircraft. Thus, some events may have included field inspections to ensure safe operations while inductions were postponed due to capacity constraints; others may have included PDM tasks that could be safely and effectively performed in the field.

Figure 3.1
Shifting PDM Facilities Caused a Temporary KC-135 Availability Shortfall from 1998 Through 2001

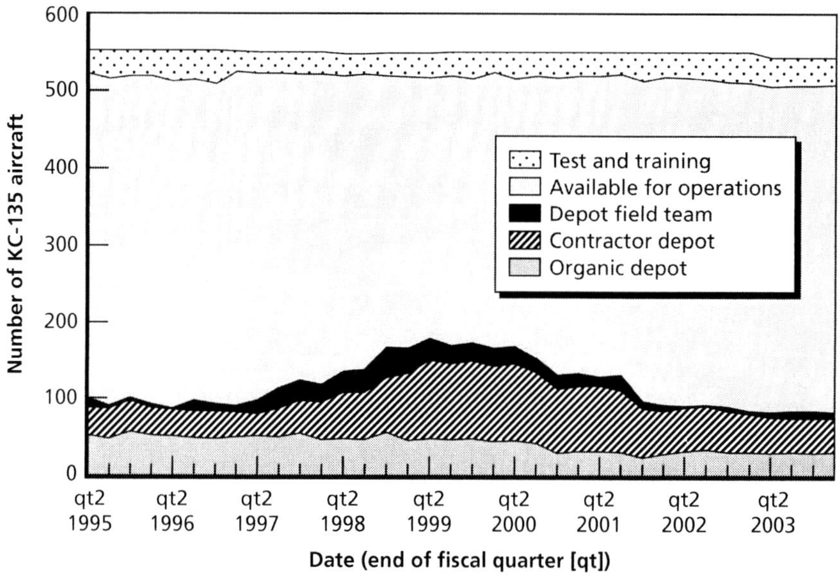

RAND *MG519-3.1*

facilities possessed 176 aircraft undergoing maintenance or modification in the depots or in the field. As the three facilities began to stabilize, the total number of aircraft undergoing PDM slowed slightly in 1999 and then diminished over the next two years to levels lower than before the transition.

Obtaining Relevant Data

To use the PDMCAT model, an analyst will need to know or estimate

- the forecast depot product standard hours (DPSH) per PDM over time
- minimum hands-on flow time, which can be estimated as total DPSH per PDM/number of DPSH that the depot can accom-

plish on one aircraft in one day, if there were only one aircraft in the facility (DPSH/labor application rate)
- capacity in number of docks over time
- PDM interval in years
- number of aircraft in work at the beginning of the forecast period
- historical annual PDM production of aircraft for the period of one PDM interval before the forecast period
- forecast of aircraft in the force over time, accounting for attrition, retirement, and acquisitions.

From these values, the PDMCAT model performs calculations that forecast two series of values:

- the number of aircraft in work at the end of each future year
- the number of aircraft produced in each future year.

Some parameters may be observed directly or are known in advance, for example, the historical PDM production counts and the number of aircraft in work at the beginning of the analysis. Other key parameters, such as future capacity, future fleet size, and PDM interval, may be observed directly at the outset; but future years' values may change, depending on the scenario being analyzed.

The two remaining parameters, workload and minimum hands-on flow time, require more-extensive analysis to estimate. Internally, the model uses the minimum hands-on flow time (R_0) that the system could achieve if there were only one aircraft in the system. (This minimum flow time cannot be achieved in practice, because it would require having personnel and other resources standing idle until the one aircraft in the system reached the particular stage at which those resources could be applied. The result, while very fast for that one test aircraft, would also be very inefficient and costly compared to a more fully loaded system.)

Therefore, the minimum hands-on flow time parameter (R_0) cannot be observed directly. It could be observed only if one could stop normal operations of the PDM facility, then send a reasonable sample of aircraft through PDM one at a time. One way to estimate R_0 is to use the labor application rate in conjunction with the forecast workload. The labor application rate is a measure of the number of hours of productive labor that can be applied (or "earned," as in the earned DPSH) on a single aircraft in a day. For example, a labor application rate of 250 hours per aircraft day implies that 250 hours worth of labor could be applied to the production of one aircraft in a workday. A dock within a facility may have 300 hours of labor available on a given day, but only 250 hours can be applied directly to production on one aircraft; the remaining 50 hours may be expended awaiting access because of competing task requirements, waiting for shared equipment, waiting for a preparation task to enable another task, etc. It does not include leave, training, shop cleanup, test-equipment adjustment, or the myriad of other shop tasks that are required to keep the shop productive. Depot managers are likely to have a good estimate of the labor application rate (sometimes called the hands-on burn rate) for the aircraft that their facility serves. If the modeler has a reliable estimate of the hands-on burn rate, then an estimate of R_0 is given by dividing the DPSH by the hands-on burn rate.

In this study, we did not have an estimate of either the single-aircraft flow time or a hands-on burn rate. For this reason, we derived R_0 using historical data; the details of this derivation are presented in Appendix C. The general idea for estimating the labor application rate is to use the change in the number of aircraft in PDM status (from one period to the next), the capacity (measured by the number of docks), and the annual production to estimate what the recent minimum hands-on flow time has been, then use that estimate and the average DPSH or depot product earned hours (DPEH) produced per completed PDM over that period.

Of course, that value of R_0 is valid only for the workload that was then being applied to aircraft in the facility. If that workload were to change, one would anticipate that the R_0 would also change accordingly. Thus, it is important to estimate the maximum labor application rate and use that value to estimate future R_0 values as workloads change. In that way, one can estimate future throughput and work in process as the future workloads vary. To accomplish this, we estimated that the labor application rate is equal to the DPSH divided by the estimated hands-on flow time. We summarize the results of our calculations of the minimum hands-on flow time and hands-on burn rate for several years during the KC-135 PDM backlog growth and recovery period, shown in Table 3.1. The burn-rate results are computed by dividing the average DPSH for aircraft produced that year by the minimum hands-on flow time.[2]

We will also use the Table 3.1 estimates for 1997 through 2002 in Chapter Four to demonstrate how the PDMCAT calculations work. In particular, we will demonstrate how the changing hands-on flow times affect the number of aircraft available for operations.

The computations of minimum hands-on time in Table 3.1 measure the effect of some other, deeper, underlying change. We identified four changes that may have contributed to the diminishing times starting in 1999 and continuing through 2005:

1. steady improvement in the two contractors' processes
2. process improvements introduced at OC-ALC in 2000–2001
3. increase in the labor available
4. reduction in concurrent modifications in 2001 and beyond.

It appears that all four changes contributed to the organic and contractor depots' success in reducing the work in process. For a more extensive discussion of this topic, see Appendix C.

Estimating Future Workloads

We found three different analyses that yielded different predictions of future KC-135 work per PDM. The first, an analysis focused solely on the KC-135, was the *KC-135 Economic Service Life Study* (ESLS) (Sperry et al., 2001). That study used regression analyses combined with interviews of

---

[2] We recommend using DPEH for this calculation. The DPEH value reflects both the DPSH planned for aircraft produced each year and some (usually positive) variations from the planned workload that depend on the amount of work discovered during the PDM inspection process. For this study, we obtained a history of DPEH from the 437th Tanker Sustainment Group (437 TSG). DPSH may slightly underestimate the workload if DPEH are not available.

Table 3.1
Annual Average Minimum Hands-on Flow-Time Estimates

FY	Capacity (docks)	Production	In PDM Status Oct. 1 (year start)	In PDM Status Sept. 30 (year end)	Minimum Hands-on Flow Time (days)	Labor Application Rate (DPEH/day)
1996	72	111	96	91	134.5	—
1997	72	94	91	79	152.1	194
1998	72	81	79	91	176.6	174
1999	72	73	91	120	214.2	146
2000	72	88	120	126	189.3	178
2001	72	109	126	95	146.1	234
2002	72	109	95	86	135.0	284
2003	72	113	86	77	124.2	—

NOTE: In-work numbers include aircraft in the OC-ALC, SM-ALC, BASC, and PEMCO facilities undergoing PDM and modification. They do not include aircraft undergoing modification at other facilities or those undergoing depot maintenance in the field.

engineering experts to develop a forecast of KC-135 PDM workload growth, which included how fatigue cracking and corrosion growth might be expected to affect workloads in future years. Those experts identified several different workload types with different underlying growth rates, as shown in Appendix C. The second workload forecast was an updated version of the ESLS workload forecast developed by officials in the KC-135 TSG (437 TSG) in 2005. The third was a purely statistical RAND Project AIR FORCE (PAF) study that sought to discover and characterize maintenance life-cycle workload patterns that were common across all Air Force fleets, rather than a pattern that may reflect some idiosyncratic, temporary behavior in a single fleet's history (Pyles, 2003). The PAF forecast equation is shown in Equation C.3, Appendix C.

The system program director's (SPD's) forecast of future PDM work closely resembles the ESLS's, but predicts slightly fewer PDM hours, as shown in Figure 3.2. While the ESLS and SPD forecast are very similar to one another, the PAF forecast grows more

Figure 3.2
The ESLS, SPD, and PAF PDM Forecasts Initially Agree, but Diverge as the Fleet Ages

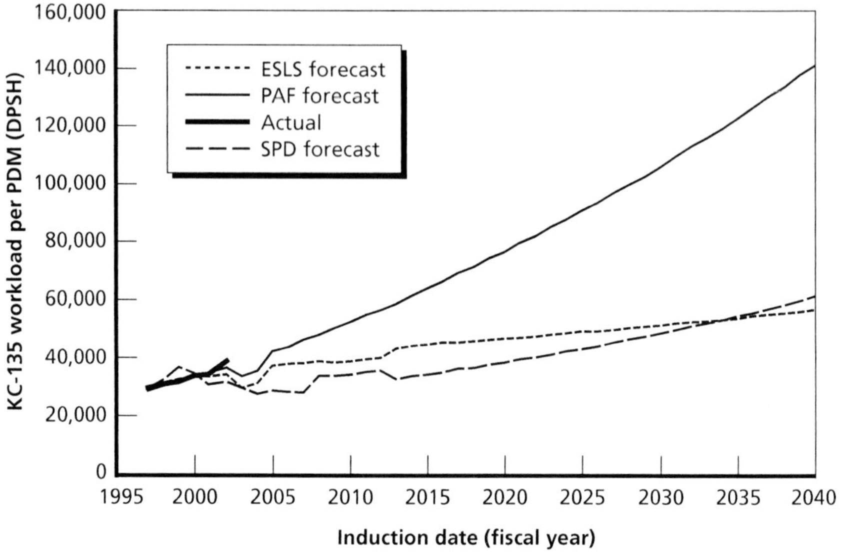

RAND *MG519-3.2*

rapidly and begins to diverge strongly before 2010.[3] The Air Force simply has not operated any aircraft fleet more than 45 years, but current plans call for operating portions of the KC-135 fleet until about age 80. The PAF forecast in the figure shows what would happen if the PDM workload pattern measured through 1998 were to continue, while the ESLS and SPD forecasts suggest a more moderate growth rate. We also show the actual workload levels through 2002 (as the heavier solid line in the figure) for comparison.[4]

---

[3] It is not our purpose here to justify the relative validity of any of these forecasts. All three come from credible analysis and data sources. Suffice it to say that there is considerable uncertainty, and even some controversy, about the future growth of the KC-135 PDM workloads and costs.

[4] Official measures of KC-135 PDM workloads in the period after 1998 did begin to level off; however, the per-PDM cost in constant dollars has continued to rise as material repairs (tracked in a different account) have increased markedly. Some workloads that might have

The 2003–2004 dips in the ESLS and PAF forecasts are due to the planned postponement of a 5,000-DPSH rewiring task. When the ESLS was conducted, a three-phase rewiring effort was already under way, but the details of the third phase had not yet been defined. Because there were no concrete plans for that effort, the ESLS team elected to exclude that task from its forecast of future PDM maintenance work. We took the liberty of adding that workload back to its forecast for 2005 and beyond, given the planned completion of the design and prototyping efforts then under way (Montgomery, 2003). The SPD's more-recent forecast reduced the third phase rewiring task to 3,500 hours and further delayed its initiation to 2008. We also adjusted the PAF workload forecast for the two-year rewiring postponement to reflect near-term workload plans. Likewise, we eliminated a topcoat repair from the ESLS forecast because the SPD has elected to address the problem through a field maintenance effort.

All forecasts provide aggregate estimates of future PDM workload growth. Based as they are on historical workload data and engineers' experience, they also reflect the gradual introduction of technology changes, capital-labor substitution, learning curves, and the gradual elimination of non–value-added tasks. If one extrapolates those relationships, one is implicitly assuming both that the underlying workload will grow and that the rate of technology change, capital-labor substitution, and process streamlining will continue as before. Because the forecasts do not agree over the long term, we examine the likely PDM work in process under the ESLS, SPD, and PAF assumptions in the next section.

Estimating Future Labor Application Rates, or Hands-on Burn Rates

Work scheduling and other technical details of the production process affect the hands-on flow time, R_0. Industrial engineers sometimes use the concept of labor-application rate (or labor-hour burn rate) to represent that relationship. The labor application rate is the amount of direct labor that can be accomplished on a single job per day.

---

been recorded as labor in earlier years may have been shifted to material expenditures, changing the scope and meaning of the current workload measures.

In essence, the labor application rate is the labor capacity to do work, given the current facilities and processes. Table 3.2 summarizes the interactions between labor application rate and R_0.

One can measure an aggregate, average hands-on labor application rate across the entire KC-135 PDM process without analyzing all the detailed task flow times and their interacting constraints.[5] Each PDM facility records the amount of work "earned" for each aircraft as it undergoes PDM and modification.[6] Of course, the maintenance processes and the recording processes vary across the three locations, so the cross-facility measures are not commensurate.[7] To overcome that disparity, we used the workload-earned data from the organic facility, recorded in the G072 data system,[8] to calculate the average amount of work performed per PDM each year and assumed that the contractor facilities accomplished the same work, though the reported labor hours may vary. When the average amount of labor earned (i.e., DPEH) is

---

5 By *interacting constraints,* we mean task sequencing, safety, and "elbow room" conflicts that prevent undertaking multiple tasks at the same time. Some tasks during certain stages of the process, such as inspection, may be conducted without much mutual interference, but others, such as system testing or defueling, require that no other tasks be performed simultaneously. Our aggregate hands-on burn rate is intended to be an average of how much work could be done under ideal conditions on a typical workday on a single aircraft.

6 *Earned hours* in the organic PDM shops means the number of hours charged to the customer for a particular task. It does not include time when the technician was engaged in other duties (such as training or test equipment repair or adjustment), time spent waiting because of "interacting constraints," or time spent waiting for work. It also does not include time required to correct an incorrectly performed maintenance action. Thus, it is a measure of the credit the facility takes (and the price it charges) for the work performed for a given task, not the actual time required to perform that task. Often, the hours earned are based on the standard hours allotted during planning for the task. The standard hours for tasks are renegotiated if the facility discovers an imbalance between actual and standard task times or occurrence factors.

7 For example, one facility may subcontract the repair of an item that another repairs. The first facility would record fewer labor hours but higher material costs for a given maintenance task than would the second facility.

8 AFMC's standard G072 data system is used to record the labor and material costs associated with maintenance throughout all the command's organic labor shops, including PDM shops. As PDM task completions are recorded, credit is taken for the work performed and debited against the obligation authority associated with the specific PDM aircraft where the work was performed.

Table 3.2
Relationship Between Labor Application Rate and R_0

Management Action	Change in Labor Application Rate	Change in R_0
Add shift work or overtime	Increase	Decrease
Process improvement or improved technology	Increase	Decrease
Capital labor substitution	Decrease	Decrease

divided by the estimated minimum hands-on flow times from Table 3.1, we can then compute the labor hours that the facility can apply per hands-on day, or the labor application rate.

As shown in Figure 3.3, the fluctuations in the labor application rate (dashed line) mirror the fluctuations in hands-on flow time (solid line) during the period of 1997 through 2002. It appears that one major contributor to the increased flow times from FY 1997 through FY 1999 was a reduced ability to apply labor to aircraft in PDM status. Even though Figure 3.2 shows that the workload per PDM continued to grow through 2002, Figure 3.3 shows that by 2000, the three facilities were able to increase their composite labor application rate enough to reduce the average hands-on flow time despite the larger workloads.

Those improved labor application rates may reflect two different underlying factors: increased labor and improved scheduling efficiencies. If additional labor resources become available, schedulers may be able to add work shifts or increase average manning in each dock so that they can apply more labor per day to each aircraft. More interestingly, they may also be able to improve the labor application rate without additional labor—by finding new opportunities to perform multiple tasks simultaneously.[9]

---

[9] We note in passing that some process changes, such as automating a task, may also reduce the hands-on burn rate by replacing labor with capital. That change may or may not affect the hands-on flow time within a stage. Today's PDM process continues to depend heavily on direct labor. Were that to change, new, lower labor application rates would emerge that reflect the reduced labor requirement and the changed hands-on flow times.

Figure 3.3
Relationship Between Labor Application Rate and Flow Time

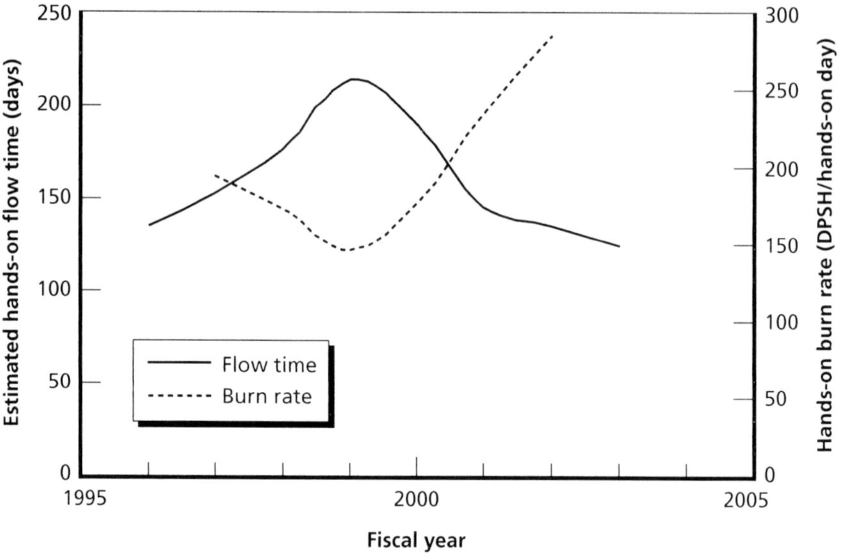

RAND *MG519-3.3*

Scheduling efficiencies at all three facilities may have been a major contributor to the labor application rate improvements after 1999, as the BASC schedulers gained experience, the PEMCO facility overcame its labor difficulties, and the OC-ALC facility introduced process-flow improvements. On the other hand, all three facilities also experienced improved labor availability, as BASC grew from essentially zero capacity, PEMCO overcame its labor problems, and OC-ALC's KC-135 shop increased its labor pool.

Of course, the mechanism that enabled those improvements is important because it may offer some insight into possible future improvements. Currently, one cannot determine how much either factor (additional labor or improved work scheduling) contributed to the improved labor application rates and flow times. As a consequence, we will assume in this analysis that the latest measured labor application rate (taken in 2002) is representative of labor application rates that the three facilities will be able to achieve in the future.

We note that Figure 3.3 displays a sharp upward trend in the labor application rate over the four years from 1999 through 2002. If that upward trend were to continue, even at a more deliberate pace, the three facilities might be able to continue improving their process flow times in the future. If so, the forecasts in this monograph would be pessimistic, estimating a larger number of aircraft in work than would actually emerge.

Near-Term Planning: Why Recent Production Matters

The KC-135 fleet PDM process has just experienced a turbulent period during which previously stable flow times and production rates were disrupted by a period of low production outputs, followed by a period of higher-than-usual production outputs. If the KC-135 PDM managers were to follow the AFTO 00-25-4 interval prescriptions slavishly, those production fluctuations would reappear as induction fluctuations, initially creating a temporary reduction in inductions from 2003 through 2005, followed by a temporary increase in inductions from 2006 through 2008, as shown in Figure 3.4.

Of course, those induction fluctuations would also cause fluctuations in the total amount of KC-135 PDM work required each year during that period, which would probably cause cost-conscious PDM managers to lay off skilled workers in 2004 and 2005 and then attempt to hire them back in 2006 and 2007.[10] To overcome that turbulence, the managers could also induct a few aircraft early, reducing the size of the layoff and that of the backlog that would be generated by the subsequent peak.

Indeed, the KC-135 SPD and the lead major air command, AMC, directed the PDM facilities to induct an additional 24 aircraft beyond the level indicated by AFTO 00-25-4 in FY 2003. That required inducting some aircraft that would normally have been inducted in 2004. Thus, if the SPD were to return to the strict AFTO 00-25-4 60-month induction interval in 2004, the number of

---

[10] The actual layoffs might be delayed up to a year after the induction rate declined because much of the work each year consists of aircraft carried over from the previous year. Indeed, BASC laid off about 100 personnel from its San Antonio KC-135 PDM facility early in 2004, and OO-ALC reassigned some of its KC-135 personnel to other workloads.

Figure 3.4
AFTO 00-25-4 Induction Rules Would Cause KC-135 Workload Troughs and Surges

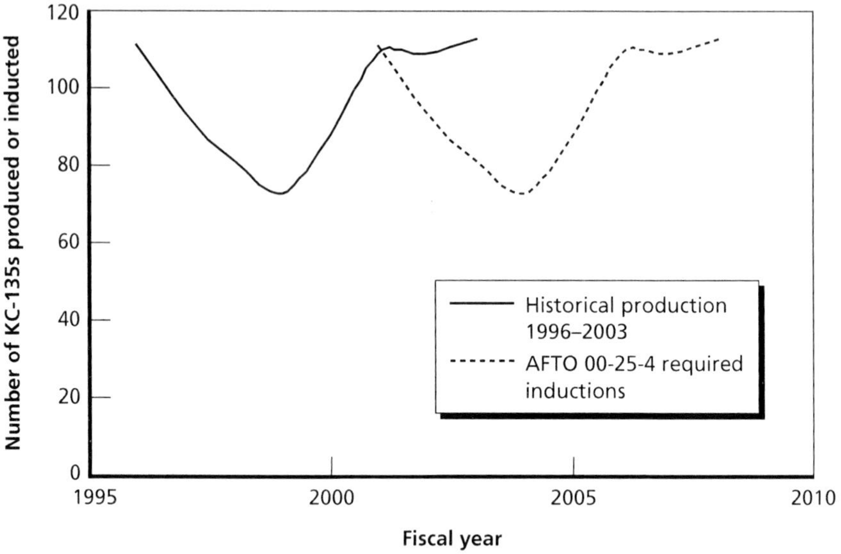

RAND *MG519-3.4*

inductions that were due would drop even more sharply than otherwise because many of the aircraft that should have waited until 2004 already would have been inducted in 2003. Thus, the SPD and AMC also directed the PDM facilities to induct additional aircraft in subsequent years in such a way as to smooth out the workload for those few years.

In Chapter Four, we examine what would have happened if that policy had *not* been followed.

Designing Cases

Of course, models such as PDMCAT are designed to evaluate different assumptions about future operations by comparing the results of alternative cases. Those cases' features will vary, depending on the issues being addressed.

We designed PDMCAT to address a subset of the full range of issues PDM maintenance planners and users face. Broadly speaking, we envision the model being used to assess

1. workload plans
2. PDM capacity plans
3. PDM interval changes
4. PDM burn-rate changes
5. changes in fleet size.

It may be helpful to indicate what we leave to other analytic methodologies:

1. detailed facility design
2. capital-labor substitution
3. skill-mix design
4. process design and improvement
5. technology enhancement
6. per-aircraft workload estimation.

To be sure, these other, more-detailed factors can affect the labor application rate or the workload required in the PDM facility, so it would be possible to evaluate them if their effects were translated accordingly. However, additional analytic methods would need to estimate adjusted factors for PDMCAT.

Against that backdrop, we now turn to our sample analyses; the findings will be discussed in Chapters Four and Five. In Chapter Four, we present the results of three separate comparisons: of the model's forecasts against the 1997–2003 backlog surge and recovery, of near-term forecasts with and without the accelerated inductions planned by 437 TSG, and of long-term aircraft work-in-process forecasts based on the ESLS and PAF workload forecasts. For the last forecast, we examine how changing the PDM capacity or the burn rate might change the outcomes. In Chapter Six, we conduct a separate excursion to examine the potential effects of retiring part of the KC-135 fleet.

First, we outline the analytic approach for the 1997–2003 historical comparison, then for the accelerated inductions, and finally for the long-term strategic plan.

Comparing PDMCAT Forecasts Against Recent History

In this demonstration of the PDMCAT process, we used the actual reported workloads from the organic PDM facility at OC-ALC, along with the annual burn rates in Table 3.1, the historical dock capacity, and the actual inductions in FYs 1997–2001 to estimate production counts and aircraft in work over the 1996–2003 period.[11] While the demonstration is reflexive for the period from 1997 through 2001 because the parameters in Table 3.1 are based on the same data, the forecast results for 2002 and 2003 are not reflexive. Thus, it provides some demonstration of the model's ability to forecast actual aircraft in process and production.

Near-Term Prediction: Leveling Workload Fluctuations

The PDM backlog in the 1997–2002 period led to a potential fluctuation in future inductions, as discussed earlier. Indeed, the relatively poor match we found between the initial 2003 prediction and the actual aircraft in work was due to the SPD's early induction plan.

To demonstrate how earlier inductions could smooth the total workload and aircraft-in-process levels, we used the PDMCAT forecasts of inductions, production, and work in process to develop an induction program that would level out the number of aircraft in PDM status for the period of 2003–2005. This was an iterative analysis, in which we initially implemented the AFTO 00-25-4 rules strictly, then adjusted the annual interval so that we would induct more or fewer aircraft as needed to maintain a steady workload over the next several years.

---

[11] Of course, this is not a comprehensive validity test of the PDMCAT model because we used the underlying mathematics associated with PDMCAT to calculate the values that would yield a matching result for the first few years. However, the 2005 forecast does match actual experience quite well, once the inductions were corrected to match the 437 TSG's modified induction policy.

Strategic Planning: Planning for the Unknown

Finally, we assessed the current KC-135 PDM physical and labor capacities against two different long-term workload forecasts. As described earlier, the ESLS forecast projected a considerably more optimistic workload level than did the PAF KC-135 forecast. If the ESLS forecast were to occur, the current facilities' capacity may generate much more acceptable aircraft availability levels than they can with the PAF workloads. Thus, our first long-term forecasting example in Chapter Four compares the aircraft in work under the two different workload forecasts.

We then turn to evaluate actions the Air Force or the repair facilities could take to improve those forecasts' availability outcomes. First, we increase capacity by adding 50 percent more physical facilities and continuing to operate them in the same way (i.e., with the same processes and technology, yielding about the same hands-on burn rate per aircraft per day). Then we increase the burn rate by 50 percent, without increasing the physical capacity.

To round out these analyses, Chapter Four concludes by comparing the two solutions' potential financial requirements to the base cases. While reducing the average PDM flow time (by either increased capacity or increased burn rate) will reduce the number of aircraft in PDM, it also causes those same aircraft to reenter the process earlier in subsequent years—thereby increasing the total required obligation authority in later years. We examine the size of that trade-off between making additional aircraft available and meeting annual financial requirements.

Strategic Planning: Force Restructuring

In Chapter Five, we consider AMCs' plan to reduce the KC-135 fleet by retiring KC-135E aircraft until a fleet size of 490 aircraft is reached. Once again, we evaluate the PDM outcomes against both the PAF forecast and the SPD forecast, assuming the current PDM capacity. We also evaluate the outcomes if the capacity were increased in proportion to the per-PDM workload.

CHAPTER FOUR

Findings

This chapter begins by using the 1997–2003 KC-135 hands-on flow time estimates and inductions from Chapter Three to illustrate how the model uses them (along with capacity, projected inductions, and initial work in process) to estimate future work in process, production, and inductions. We first seek to replicate recent historical experience, demonstrating the model's ability to generate plausible, realistic forecasts.[1] Indeed, the illustration demonstrates the model's sensitivity to the induction policy, showing that the 2002 and 2003 inductions were accelerated with a substantial effect on the number of aircraft in PDM status. The illustration then follows that problem out a few years in a near-term forecast that demonstrates how the model can be used to moderate future fluctuations in the number of aircraft in PDM status.

We then turn to a demonstration of the model's use in a longer-term forecast. While the near-term forecast illustrated the model's usefulness in adjusting PDM induction policies to manage near-term (one- to five-year) availability, the longer forecasts address a 30-year horizon, suitable for evaluating alternative physical capacity plans, PDM process improvements, or fleet replacement programs. For the near-term plans, we assume that facility capacities are relatively inflexible but that some near-term PDM induction policies could be executed. The long-term forecasts consider how more-extensive changes in physical capac-

[1] We emphasize again that this is not a comprehensive validity test, though it does give us some confidence that it can reflect realistic outcomes, given reasonably accurate estimates of the PDM induction policy, PDM capacity in docks, recent years' production history, projected workloads per PDM, and initial aircraft in PDM status.

ity or the maintenance process might affect availability. The long-term forecasts consider the implications of two different workload forecasts (ESLS and PAF) for future depot capacity requirements. Finally, we use the model's estimates of inductions with the recent PDM workload forecasts to estimate future long-term budget requirements.

Estimated KC-135 Work in Process and Historical Values

From 1996 to 1999, the number of KC-135 aircraft in PDM status grew significantly, leaving fewer KC-135 tankers available for operations and training. After that period, the three facilities improved their joint production levels and reduced the number of aircraft in work to below 1996 levels.

We used PDMCAT and the estimated annual average hands-on flow times for 1997–2002 from Table 3.1, the estimated capacity, the number of aircraft in work at the end of 1996, and the annual inductions from 1997–2003 to forecast the annual production and remaining aircraft in work at the end of each fiscal year from 1997 to 2003. For 2003, using the 2002 hands-on burn rate to estimate the hands-on flow time with the PAF-projected workload allowed us to compare the hands-on flow-time calculation based on the burn rate (119 days) to the measured value in Table 3.1 (124.2 days, a difference of about 4 percent). We then compared these estimates to the actual number of aircraft in work at the end of each year to PDMCAT's prediction.[2]

Comparing Forecast to Actual Aircraft in PDM Status

Figure 4.1 shows the results of applying the model to the data available from 1996 to 2003 by using the standard 60-month PDM interval specified in AFTO 00-25-4. The solid black curve shows the number

[2] We do not view this comparison as a validity test of the model, only an illustration of how changes in the hands-on time can affect the production and remaining aircraft in work. Only the FY 2003 value for the hands-on rate was estimated from the predicted workloads and measured burn rates. The other values were generated algebraically from the aircraft-in-work data.

of aircraft in PDM status at an organic or contractor facility during that time, as reported in the Air Force Knowledge System (AFKS).[3] The dashed black curve shows the PDMCAT estimates.

From the end of 1996 through 2001, we forced the model inductions to match the actual inductions experienced by the PDM facilities during that period. After that, we allowed the model to induct the aircraft that had been produced five years earlier. In the period before 2002, the errors are small because the model used (1) the measured values of the hands-on time, (2) the exact number of aircraft in work at the end of 1996, and (3) the exact number of inductions from 1997 through 2001. In 2002 and beyond, we allowed the model to induct the number of aircraft produced

Figure 4.1
Actual Versus Forecast KC-135 Aircraft in PDM Status: AFTO 00-25-4 Induction Rules

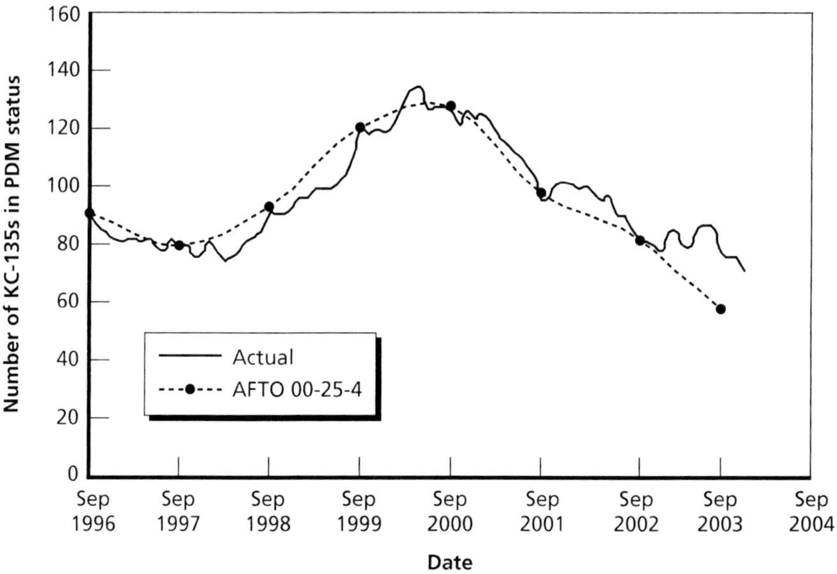

RAND *MG519-4.1*

---

[3] We did not count aircraft that were in the depot or contractor facility for fewer than 125 days as aircraft in PDM status. Also, we did not count aircraft that were at the depot but not yet inducted as aircraft in PDM status.

five years earlier. In 2003 and beyond, we allowed it to use its own estimate of the hands-on flow time, based on the 2002 labor application rate and the PAF workload forecast.

Note that the PDMCAT work-in-process forecast underestimated the actual value in 2002 and 2003. That is, the model predicted that fewer aircraft would remain in work than actually remained in that status. In general, such an optimistic prediction could have occurred for any of three reasons:

1. The model's workload forecast was too low.[4]
2. The model's labor application rate was too high.
3. The depot inducted more aircraft than the model predicted.

In this case, we found that the errors in 2002 and 2003 were caused by higher inductions in those years. That was clearly the case in 2002. We used the measured, not the forecast, hands-on flow times and thus a method not dependent on the workload forecast and the labor application rate. When we compared the AFTO 00-25-4 forecast against the actual inductions, we found that the depot inducted five more aircraft than required by AFTO 00-25-4 in 2002. In 2003, it inducted 28 more.[5] After adjusting the 2003 induction policy to match the actual induction values, PDMCAT computed 2002–2003 estimates of aircraft in PDM status that matched the actual experience much more closely, as shown in Figure 4.2.

Computing Production and Future Induction Values

Of course, the additional aircraft inductions led to both to a higher level of work being in process at the end of 2003 and a higher number of air-

4 The PAF workload forecast includes all categories of work except modifications. It thus covers the basic PDM package, economy tasks, and tasks undertaken to remedy material conditions that required labor over and above the labor included in the basic PDM package.

5 Discussions with 437 TSG and the OC-ALC production facility confirmed that aircraft had been inducted early in both 2002 and 2003. We achieved the required induction levels by reducing the interval by about one week in 2002 and by about three and one-half months in 2003.

Figure 4.2
Actual Versus Forecast KC-135 Aircraft in PDM Status: Adjusted Induction Rules and Rewiring Workload

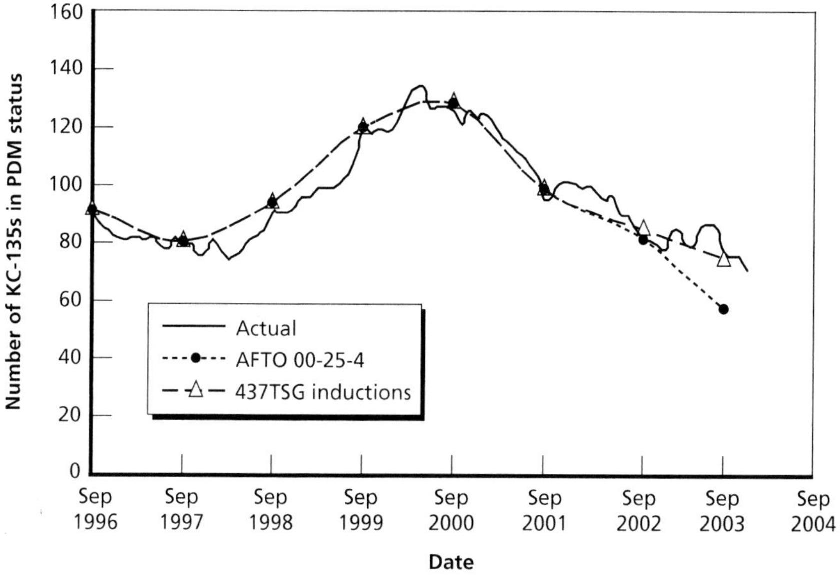

RAND *MG519-4.2*

craft being produced. Figure 4.3 shows the PDMCAT estimate of the annual number of aircraft produced through 2003, compared to both the AFTO 00-25-4 policy and the actual production. If one were to follow the AFTO 00-25-4 five-year rules for the KC-135 precisely, the aircraft produced in 1999–2003 would be the aircraft inducted in 2004–2008, so the fluctuating production in 1999–2003 would normally cause an identical fluctuating induction requirement in 2004–2008, as shown by the dashed line in Figure 4.4.[6] However, the early inductions in 2002 and 2003 would have two consequences if one were to return immediately to the AFTO 00-25-4 rules, as shown by the solid line in Figure 4.4:

---

[6] Discussions with OC-ALC SPD officials confirmed that the SPD office is actively managing inductions, within the parameters of the AFTO 00-25-4 requirements, to smooth production workloads.

Figure 4.3
Accelerating Inductions Also Increased Production

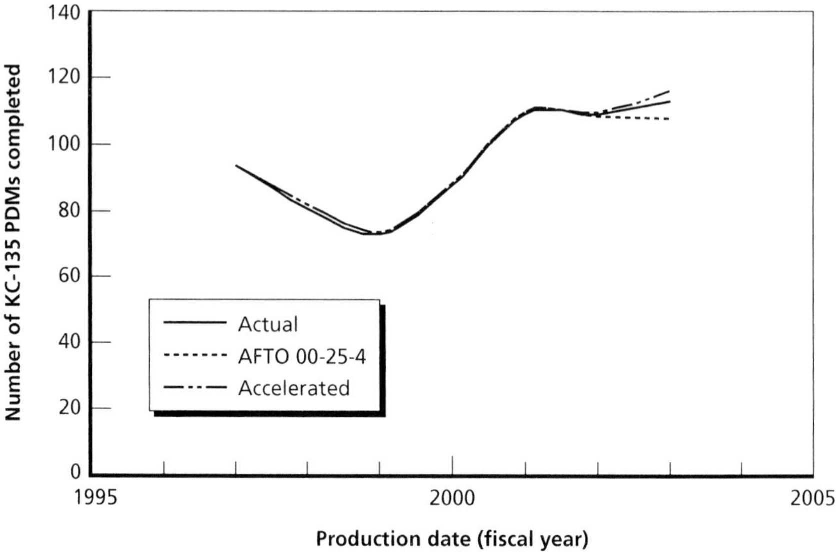

RAND *MG519-4.3*

1. Inductions in 2004 would diminish substantially.
2. Inductions in 2008 would increase slightly.

The first consequence would arise because aircraft that would have been inducted in 2004 under AFTO 00-25-4 had already been inducted in 2003. The second consequence would arise from the earlier production of the early inducted aircraft.

Forecasting and Managing Near-Term KC-135 PDM Work in Process

Of course, such a sharp reduction in annual inductions in 2004 would further reduce the number of KC-135s in PDM status well below the level achieved in 2003. The solid line in Figure 4.5 shows the number of aircraft that the PDMCAT forecasts would be in PDM status for

Figure 4.4
Early Inductions in 2003 Affect Induction Requirements in 2004 and 2008

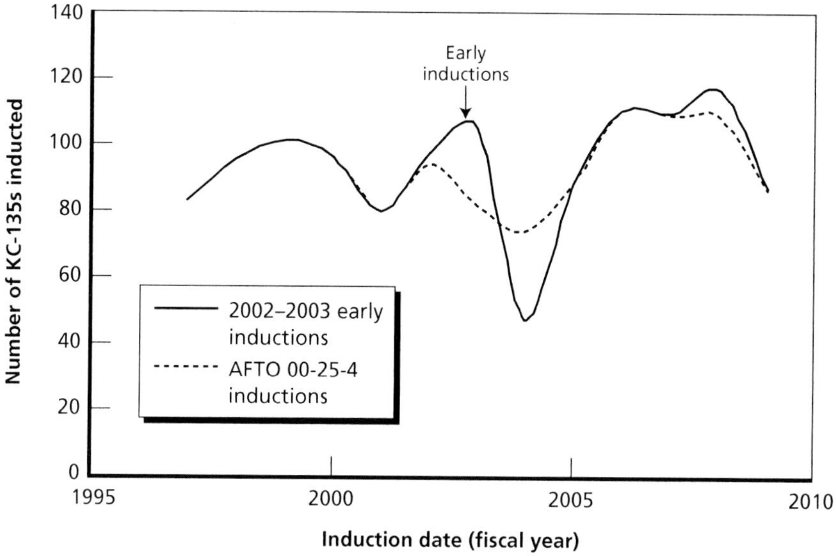

RAND *MG519-4.4*

the remainder of the decade if the AFTO 00-25-4 induction policy had been suddenly reinstituted in 2004. Under that policy, fewer than 50 aircraft would have been inducted in 2004. When combined with the already low level of work in process and the high labor burn rate recently achieved, the number of aircraft remaining in PDM status at the end of 2004 would diminish to a record low level.

Conversely, inductions from 2006 through 2009 had been poised to achieve record high levels. With the current workload forecasts, capacity, and burn rate, the levels of aircraft in PDM status in the three facilities would have been reminiscent of the 1999–2000 peak. Of course, if the actual workload were less than the forecast for the 2006–2009 period, the number of aircraft in work would also decrease.

Also of interest, but not shown in the figure, the cyclical behavior introduced by the sudden return to the AFTO 00-25-4 regimen would propagate for several decades. Each cycle would shrink compared with the previous cycle, depending on the amount of capacity available.

Figure 4.5
Accelerating KC-135 Inductions in 2004–2008 Would Improve Availability in 2006–2009

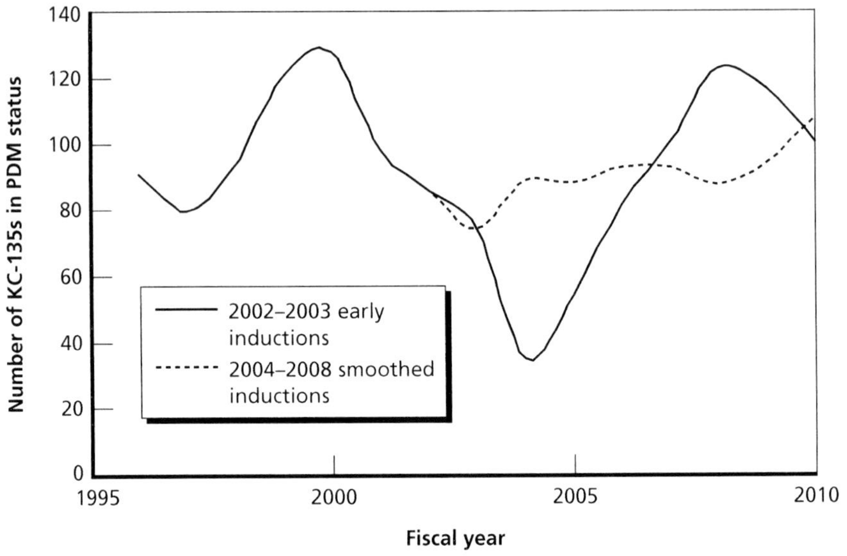

RAND *MG519-4.5*

Workload Management Can Mitigate the Near-Term Availability Shortfall

We anticipated that it would be possible to extend the SPD's policy of accelerated inductions to shift some of the peak inductions to earlier years, thereby mitigating the effect on aircraft availability. We used PDMCAT to investigate the potential for such a near-term solution. While that search process was ad hoc, we anticipate that one could use a dynamic programming approach to find a more nearly optimal solution.

One option we found is shown as the dashed line in Figure 4.5. Reducing the induction interval from 2004 through 2008 could level off the number of PDM aircraft in work to approximately 90 aircraft from 2004 through 2009.

Forecasting and Managing Long-Term KC-135 PDM Work In Process

Unfortunately, however, the solution just outlined will not solve the KC-135 availability shortfall over the long term, particularly if the workload per PDM continues to follow the late-life growth pattern discussed in the PAF aging aircraft workload patterns report (Pyles, 2003). The solid curve in Figure 4.6 shows the PDMCAT forecast for KC-135 workload growth according to the PAF forecast, assuming management of the near-term workload according to the accelerated inductions just discussed. As the figure shows, the KC-135s in PDM status would surpass the 1999–2000 peak in around 2013, assuming the current inspection interval, physical capacity, and burn rate. After that point, the number of aircraft in PDM status would continue to grow.

Perhaps surprisingly, the aircraft in PDM would not grow at an increasing rate over time. In fact, the rate of growth diminishes slightly, as shown in Figure 4.5. This is so because the PDM interval is defined as the *time between a departure from one PDM until the next PDM induction of that aircraft.* No matter how long it takes to perform PDM, the fraction of a fleet in PDM status will grow proportionately more slowly than will the PDM flow time.

There are four options for reducing the number of aircraft in PDM status after 2010:

1. Reduce the fleet size. This would result from the planned retirement of aircraft.
2. Increase the interval between PDMs.
3. Increase the physical capacity. This would entail adding docks as well as labor to man the docks at current levels.[7]
4. Increase the labor application rate. As mentioned earlier, labor application rate is the rate at which labor is applied. There are several ways to increase that rate, for example, by improving the current process, investing in labor-saving equipment, adding

---

[7] The additional capacity could be acquired by expanding current organic facilities or by adding more contractors.

shifts, working overtime, working weekends, or increasing the simultaneity of tasks.

In this chapter, we use PDMCAT to explore the last two options, increasing either the physical capacity or the labor application rate per dock by 50 percent. We increase capacity by adding docks, that is, by adding physical locations where aircraft may be parked and where work can be done. The effects of a force reduction are examined in Chapter Five.

We note in passing that either option introduces another cyclical pattern in the forecast of future workloads and availability patterns. Those patterns arise from the sudden introduction of the change in 2010, rather than a gradual increase in capacity or burn rate over time. SPDs and PDM managers could no doubt use PDMCAT to schedule

Figure 4.6
Investments in Physical or Labor Capacity Can Mitigate and Delay the Long-Term Availability Shortfall

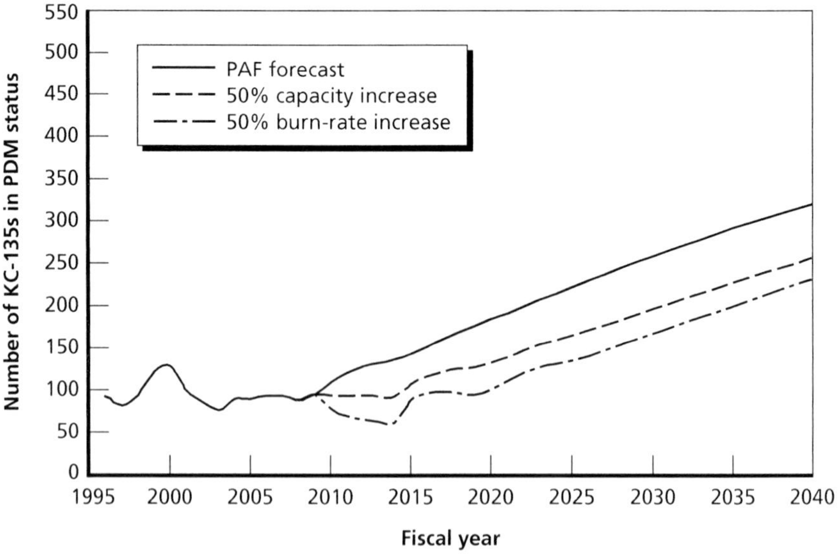

RAND *MG519-4.6*

the introduction of such changes in a way that would gain the availability benefit without introducing the cyclical pattern shown here.

When docks are added, we assume that labor is also added—to man the docks. However, the labor application rate applied to each aircraft remains the same, so the effect of this change is to decrease total flow time by decreasing the time an aircraft waits to enter a dock. That is, increasing the number of docks increases the total labor available proportionally, in effect making sure that aircraft in a dock receives work at the same rate before and after docks are added. That means there will be more docks in each stage of the PDM process, reducing the amount of time an aircraft arriving at a stage will have to wait before work can begin in that stage.

In the second case, in which we hold capacity constant and change the labor application rate, we start with the initial labor application rate (calculated using Equation C.2 to obtain an initial estimate of R_0 and dividing that estimate into the DPSH to get the burn rate).[8] We then increase the labor application rate by 50 percent, starting in 2010, thus decreasing the associated minimum hands-on flow time R_0 thereafter. The effect of this change is to reduce the amount of time each aircraft spends in each stage of PDM, reducing both the time while work is performed and queuing delays.

The two dashed lines in Figure 4.6 display the results. Increasing the labor application rate (which reduces both hands-on and queuing time) has a greater effect than does increasing the physical capacity. While increasing the physical capacity by 50 percent would delay matching the 1999–2000 peak until 2020, the same increase in burn rate would delay that event until 2024.

Each option has its limitations. Increasing the physical capacity would require displacing other workloads from similar facilities; hiring additional contractors; or acquiring additional space, buildings, and equipment. On the other hand, increasing the labor application rates so dramatically would require adding shifts, increasing labor hours per shift, or replanning the current work process—options that may have already been fully exercised.

---

8 See Appendix C.

Strategic Planning for Uncertain Future Workload Growth

The Air Force has never kept aircraft for the periods of time currently envisioned for the KC-135 fleet. Thus, there is considerable controversy over whether or not the PAF workload forecast will actually arise. Indeed, the PAF analysis of maintenance workload patterns (Pyles, 2003) suggests that there must be some limit beyond which the PDM and other workloads cannot grow. It even suggests that the PDM processes may contain some one-time repair tasks (such as the KC-135 rewiring) that could cause the workload to peak at some point, then diminish.

While the PAF analysis did not detect such a leveling in the experience of past Air Force fleets, the ESLS team estimated how such concerns might limit the rate of future PDM workload growth. When one uses their workload forecast, the number of aircraft in PDM status grows much less rapidly, as shown in Figure 4.7.

As shown on the figure, the ESLS forecast workload would lead to a smaller work-in-process growth, even without an increase in the physical capacity or burn rate. In fact, that workload would not cause the number of KC-135s in PDM status to reach the 1999–2000 peak until after 2040.

Of course, the actual workload per KC-135 PDM will emerge over time. It may or may not lie between the two forecasts we explore here. Thus, the uncertainty about what workload may emerge in the future poses a challenge for those who would advocate replacing all or part of the fleet, those who would advocate additional long-term investments in depot capacity, and those who would advocate retaining the fleet until near the end of its expected structural lifetime.[9] Next, we address how to use these forecasts to help craft a comprehensive strategy that addresses workload uncertainties.

---

[9] The ESLS engineering team estimated that only a handful of KC-135 aircraft would need to be retired for structural problems before 2040, even after considering the effects of corrosion on structural life expectancy (Sperry et al., 2001).

Figure 4.7
Forecast of Aircraft in PDM Status: Comparison Based on ESL and PAF Workloads and on a 50-Percent Labor Application Rate Increase

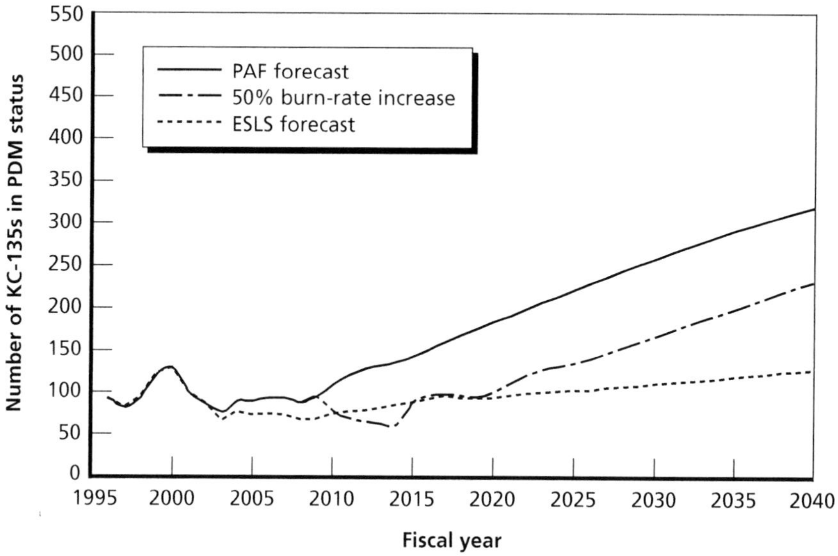

RAND *MG519-4.7*

Required Obligation Authority Depends on Workload Forecast and Management Option

PDM is funded in the MAJCOMs' annual budgets as obligations to pay for aircraft inducted each year. That is, each MAJCOM sets aside a portion of its budget each year and obligates it to the PDM facilities as each aircraft is inducted. From the viewpoint of the Air Force budget, money obligated to PDM cannot be spent elsewhere once the fiscal year begins. Thus, the authority to obligate funds for PDM is handled as an annual allocation in each MAJCOM's annual financial plans.

The PDM process is funded at a level consistent with recovering the fully burdened cost of a PDM, including a profit in the case of contractors. Thus, the PDM price paid by the MAJCOMs is intended to cover not only the labor and material costs, but also any overhead and general and administrative costs, including the depreciation of facilities and equipment. As a consequence, any investment in additional capacity by either the organic or contractor facilities would appear in the charges paid by the MAJCOMs.

Figure 4.8 shows how much annual obligation authority would be required for each of the four long-term options analyzed above. It was computed by multiplying the annual inductions PDMCAT estimated for each case by the forecast workload per PDM and \$200/DPSH.[10]

As depicted in the figure, the ESLS forecast is by far the least expensive. Of course, the reason for this is because it forecasts a much less demanding workload per PDM.

Likewise, the two options of increasing burn rate or physical capacity are more expensive than the option of continuing with the current capacity. The more expensive option is to increase the labor application rate, because it completes (and inducts) more aircraft per year and achieves a higher availability level, assuming the PAF workload forecast.

It is interesting to note that the obligation authority requirement does not increase for five years after the labor application rate or physical capacity has increased. That is because the obligation authority depends on inductions, not the completion of the PDM. Because aircraft are being inducted on a fixed five-year interval, the annual obligation authority in 2010–2014 does not change just because the facility is able to perform the work faster. However, it does change in 2015 and beyond, because the additional aircraft produced in the first half of the decade need to be inducted.

Finally, we note that, as with the long-term availability curves in Figures 4.6 and 4.7, the annual obligation authority in every case tends to level out, not grow more rapidly with age. Taken together, these figures demonstrate that neither the number of aircraft in PDM status nor the annual total fleet PDM costs grow in proportion to the forecast workload.

---

[10] The value of DPSH includes all direct labor, indirect labor, material, overhead, and general and administrative costs. The average cost of a DPSH in AFMC in 2004 was just a few cents less than \$200. In 2005, that number increased to almost \$225.

Figure 4.8
Forecast of Required Obligation Authority

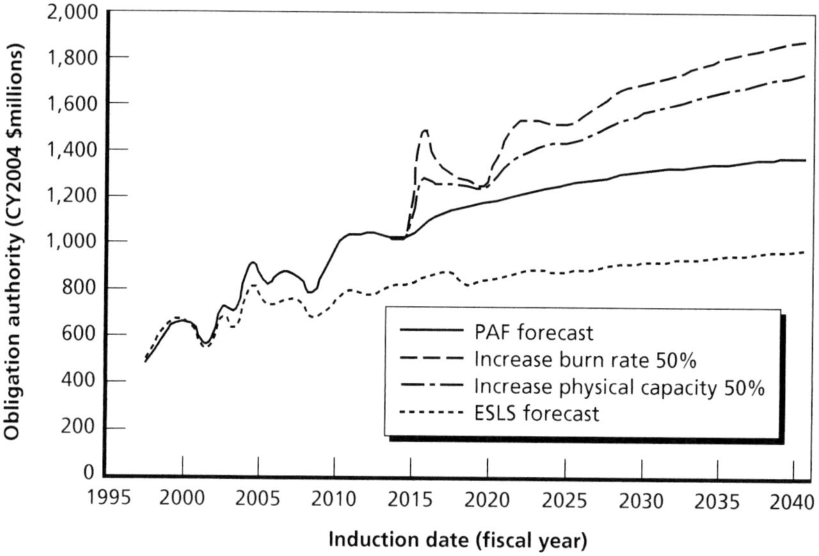

RAND *MG519-4.8*

CHAPTER FIVE

AMC Fleet-Retention Plan and Workload Forecast

In this chapter, we use the PDMCAT model to forecast future KC-135 aircraft availability based on the AMC plan to reduce the existing fleet by retiring KC-135Es until the fleet size reaches 490 aircraft. We use the workload forecast based on engineering judgment provided by 437 TSG, and we then compare the aircraft availability projection by using its workload against that produced by the statistically based PAF workload forecast. In both cases, we assume that capacity changes in proportion to changes in workload.

Table 5.1 shows the planned fleet structure. This table shows that 417 KC-135R/Ts and 73 KC-135Es will remain in the fleet. Further, it assumes that KC-135Rs are transferred from AMC, as shown in Table 5.1, to the Air National Guard (ANG) and the Air Force Reserve Command (AFRC). Current inventories in the Air Education and Training Command (AETC), U.S. Air Forces in Europe (USAFE), AFMC, and Pacific Air Forces (PACAF) were retained in place.

To initialize the model, we used the KC-135 induction, work-in-process, and production data for the 1997–2003 period. The production data were used to create a stream of future inductions for the model. As described in Chapter Three, the labor application rate appeared to stabilize during 2003 at a level near 281 DPSH per day, averaged across the three facilities. Given the per-PDM workload, one can directly compute the minimum hands-on flow time. For example, a 30,000-DPSH workload would require a minimum of 107 days without queuing, averaged across all three facilities.

Table 5.1
KC-135 Fleet Structure for Air-Refueling Analysis of Alternatives

Owner	Mission Design Series	Production Year							
		1958	1959	1960	1961	1962	1963	1964	1965
AETC	KC-135R	0	1	3	5	3	4	7	1
USAFE	KC-135R	0	1	2	2	0	5	3	2
AFRC	KC-135R	3	19	3	9	5	12	10	2
AMC	KC-135R	5	15	13	10	15	30	25	3
AMC	KC-135T	0	20	16	11	0	0	0	0
ANG	KC-135E	21	35	14	1	0	1	0	0
ANG	KC-135R	12	25	23	16	13	25	21	4
AFMC	KC-135E	0	1	0	0	0	0	0	0
AFMC	KC-135R	0	0	0	0	1	0	0	0
PACAF	KC-135R	0	0	1	0	2	1	0	1
PACAF	KC-135T	0	5	0	2	0	0	0	0

With 437 TSG engineers' moderate forecasts of PDM workloads, the PDMCAT model estimated that the future KC-135 availability would evolve as depicted in Figure 5.1.

Clearly, the number of aircraft in PDM status has already diminished substantially from the 1999 level of 176 in work described in the ESLS report (Sperry et al., 2001). This was due to four factors:

1. increased capacity created by adding a second contractor
2. process improvements introduced at OK-ALC
3. diversion of modification workloads to contractor facilities and field teams
4. temporary postponement of the final rewiring phase (more recently).

Figure 5.1
Under the KC-135 Tanker Sustainment Group Engineers' Moderate Workload Forecast, the Number of Aircraft in PDM Status Grows Slowly over the Next 60 Years

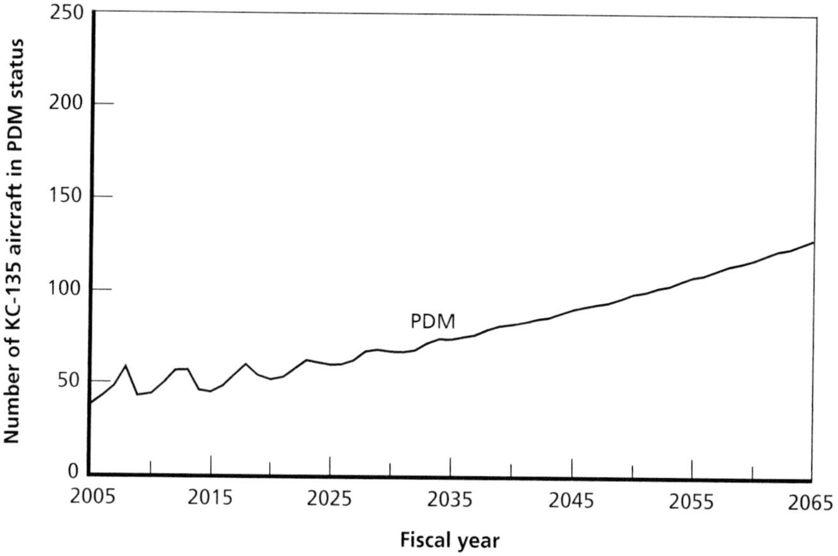

RAND *MG519-5.1*

Since the depot implemented those initiatives, the combined facilities have been able to reduce the number of aircraft in PDM status to fewer than one-third of the number in work at its peak in 1999. If the 437 TSG forecast is correct and the fleet size were reduced to only 490 aircraft, the fleet would not reach 100 aircraft in PDM status (twice the near-term forecast level) until after 2050.

Why does the model project that the number of aircraft in PDM will oscillate in the early years of the forecast? The reason is that past production surges and the planned retirements have created peaks and valleys in the fleet's induction due dates. In the real world, 437 TSG, the MAJCOMs, and the maintenance facilities would adjust inductions to avoid such cyclical behavior. As discussed in Chapter Four, we applied the AFTO 00-25-4 induction rules more strictly than would occur in the real world because the technical order permits schedule adjustments of up to six months to accommodate varying workloads or operational requirements. Because the model follows a rigid induction schedule, it takes some time before the ripples dampen out.

Figure 5.2 compares the availability forecasts of the 437 TSG workload forecast and the PAF workload forecast. In this display, we sum all the aircraft in PDM status for the two different workload forecasts.

As shown in the figure, the availability forecasts diverge strongly, with the PAF forecast projecting that the number of aircraft in modification or PDM status might reach 100 as early as 2013.

One reason for the difference is the postponement of the rewiring task in the engineers' workload forecast. The PAF forecast simply extends the pattern observed for past workloads, so it cannot reflect the near-term adjustment in plans that the SPD's forecasts embody.

On the other hand, the SPD's forecasts reflect only the validated workload known to be forthcoming. That is, the engineers must use

Figure 5.2
The PAF Workload Forecast Would Cause KC-135s in PDM to Increase More Rapidly

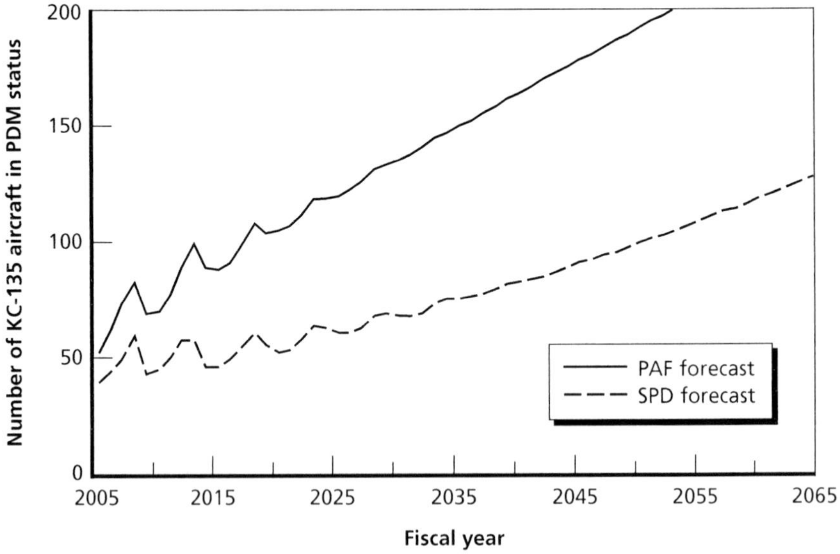

RAND *MG519-5.2*

their forecasts to compete for scarce funds to support their aircraft in the Air Force Planning, Programming, and Budgeting System (PPBS). Any workload projection they might identify must have concrete evidence that the work is needed and that the need will indeed arise. If not, the funds will be directed to other, more clearly required Air Force needs. Thus, the engineers' forecasts cannot reflect the potential workloads that may arise if major structural repairs (MSRs) or other corrosion-related workloads grow as they have in the past. Their forecasts must contain only workloads that they can directly observe in the condition of the aircraft that have recently undergone PDM. That may cause their forecasts to underestimate the future workloads that may arise either from known, but only potentially worrisome, material degradation or from heretofore-unidentified material degradation.

CHAPTER SIX

Conclusions

While we have concentrated mainly on the PDMCAT in this monograph, we used several practical examples drawn from the KC-135 fleet experience to demonstrate the model's capabilities. As the model neared completion and these sample analyses were completed, it was adopted to support the KC-135 Analysis of Alternatives (Kennedy et al., 2006). That study used PDMCAT with updated and refined KC-135 sustainment scenarios and more-recent and -authoritative data, so we do not presume to make any recommendations about that fleet's future sustainment plans. Thus, this chapter concentrates solely on our use of PDMCAT as a way to analyze future PDM plans and policies.

Observations and Conclusions About PDMCAT

From this experience, we conclude that the model can be usefully applied to PDM workload management issues and to support both near- and long-term planning of gross physical capacity, gross labor requirements, and planned process improvements for programmed depot maintenance. Given accurate estimates of future workloads, measured or estimated labor-application rates, and the current number of aircraft in PDM, the model was able to compute the number of aircraft in work accurately for several years running. It was also possible to compute aircraft in PDM status with limited loss of accuracy when the hands-on time was estimated from the forecast workload and the most recently measured labor-application rate.

While computational accuracy is important, so is the ability of the model to compute the time-varying consequences of alternative fleet and PDM facility management decisions. We were able to use the model to adjust the PDM induction interval, both to reflect a historical induction decision and to evaluate an option that would stabilize the number of aircraft in work for most of this decade. We were able to forecast a potential availability problem that could emerge in the next decade and to evaluate how changes in physical capacity, fleet size, labor availability, or labor application rates might mitigate that problem. We were also able to evaluate alternative workload forecasts and their implications for aircraft availability. While not strictly a computation from PDMCAT, its estimates of the number of aircraft inductions and the forecast PDM workload enabled us to estimate the time-varying obligation authority requirement for various options.

We envision the model being used mainly for two broad purposes: to support fleet PDM planning, programming, and budgeting and to support facility and process improvement. Given some recent PDM production history and forecasts of future work per PDM, AFMC's aircraft sustainment wings will be able to estimate and adjust future PDM inductions, production levels, aircraft in PDM status, and obligation authority more accurately for the entire Future Years Defense Plan and beyond. PDM facility managers will be able to use the detailed information about their own processes and resourcing plans (i.e., how planned facility adjustments would affect the number of docks or how planned labor availability or process changes would affect the labor-application rate) with PDMCAT to estimate how those changes would affect PDM flow times and aircraft in PDM status.

We can also imagine the model or its outputs being used somewhat less frequently to widen the choices available to Air Force planners and programmers when considering the replacement of current aircraft fleets. PAF staff have extended an earlier optimal fleet replacement model (Keating and Dixon, 2003) to include options for increasing repair capacity, perhaps changing the point at which it becomes cheaper to replace an aircraft (Keating, Snyder, and Loredo, 2004). PDMCAT provides the analytic foundation for such an investment trade-off.

Limitations of the PDMCAT Model

PDMCAT is a macro forecasting tool, and, as with all forecasting tools, the accuracy of the forecast will depend on how accurately the underlying factors have been measured. Three critical factors are required for PDMCAT:

1. *A forecast of future workloads.* Future PDM workloads are the subject of much debate. Pyles (2003) found a general cross-fleet pattern for PDM growth as fleets age, as well as a significant second-order term related to age. In an analysis focused solely on the KC-135, the ESLS used engineering knowledge and judgment and projected continued growth on KC-135 PDM workloads, although at a less pronounced rate than that predicted by Pyles. The KC-135 SPD office provided a workload forecast that closely mirrors the ESLS forecast in terms of rate of growth, but it projects fewer hours per PDM. The PDMCAT forecast of aircraft in work will vary depending on the workload forecast used. Workloads have grown in recent years; however, this fact is not conclusive evidence that the trend will continue into the future. Some argue that workload growth will necessarily taper off as all or most of the key components on the KC-135 are repaired or replaced. Available workload forecasts are based on experience with aircraft less than 40 years of age. Future experience with older aircraft may improve the understanding of how age-related workloads grow. It is therefore important to regularly review and adjust the workload forecasting equations to ensure that forecasts are as accurate as possible.
2. *An estimate of the labor-application rate.* The rate at which labor can be applied to PDM workloads may change as process improvements, learning, and technology improve the efficiency of depot personnel. For example, if the modeler uses an estimate of the most recent labor-application rate in projecting all future aircraft in PDM status and if process improvements allow the actual rate to increase, the model's prediction error will increase. It is incumbent on the analyst to ensure that the

model reflects changes in application rate. In fact, one potential use for the model is to test the cost-effectiveness of proposed process changes by observing how model outputs change as a consequence of improvements in burn rate.

3. *An estimate of depot capacity.* The PDMCAT model measures depot capacity in docks. The modeler has the option of entering a constant number of docks or of increasing the number of available docks over time. However, the PDMCAT model does not assess how the addition of docks may change the labor skill mix and affect the burn rate, nor does it address how additional docks are added. PDMCAT does not differentiate between the addition of docks within an existing facility (by freeing up space currently occupied by other workloads) from the addition of docks by hiring contractors. PDMCAT assumes that depot management will seek to maximize the effective use of their facilities by shifting or adding capacity as needed to maintain a balanced workflow and to maximize throughout.

Next Steps for PDMCAT Modeling and Use

Several PDMCAT extensions and improvements are already under way. For a recent analysis of the C-5 PDM process, we extended the model to process multiple PDM job streams with different total work requirements. (The C-5A requires both shorter intervals and more work per PDM than does the C-5B.) For an analysis of the projected KC-767 fleet with smaller, more-frequent C-checks,[1] we adjusted the model to use a smaller step size internally to improve overall accuracy.

Other refinements or enhancements are needed to improve the usefulness of the model still further. The most useful enhancement would be to develop projective (rather than historical) techniques to

---

[1] C-checks are PDM-like inspections for commercial aircraft that sequentially inspect different portions of an aircraft. In aggregate, they perform a complete inspection of the aircraft that is similar to a PDM, but the aircraft enters the C-check process more often and the amount of work per C-check is smaller.

estimate how the labor-application rate would change due to some proposed improvement in scheduling or due to increased labor availability. The technique we used for this demonstration was to measure past burn-rate performance and assume that it would never change. One has only to look at Figure C.1 to notice that the labor-application rate need not be constant. What is needed is a way for PDM facility managers to estimate how changes in their process flows or task labor requirements would change their overall burn rates. Managers already use Program Evaluation and Review Technique (PERT)–like PDM task networks to discover ways to increase the number of tasks that can be performed simultaneously.[2] It would be useful if it were also possible to use the networks to estimate how much labor could be applied daily during a typical PDM, then compute an average theoretical maximum labor-application rate, as adjusted by the amount of labor and shifts available. With such a computational process, PDM facility managers could rapidly identify and evaluate process changes or labor availability levels that might reduce both overall flow times and work in process.

Finally, the model needs to be placed into a format that is more accessible than the spreadsheet form currently used. While that format was adequate for our prototyping and exploration, data entry was error prone, and cross-case comparisons were laborious. Now that the early prototyping phase is complete, the model needs to be transferred into a more-traditional format.

---

[2] PERT is designed to display and monitor the progress of complex projects, such as an aircraft PDM. It consists of a detailed network of interdependent tasks with defined maximum and minimum durations. By rearranging the interdependencies in the network (i.e., which prior tasks must be completed before a given task can start), it is possible to increase or decrease the number of tasks that can be performed simultaneously. That number, of course, would increase or decrease the labor-application rate.

APPENDIX A

Different Approaches to Forecasting Availability

In this appendix, we present a review of previous approaches to modeling aircraft availability and some general theory on queuing network analysis. We introduce the BJB approach to analyzing queuing networks. We also discuss the interaction between workload growth and PDM capacity.

Open Versus Closed Processes

As Figure 2.1 showed, the total operations/PDM process is a closed system, with no aircraft entering or leaving except when they retire or are otherwise removed from service. These aircraft circulate around this closed network all their operational lives, which leads to a conservation of matter, whereby the aircraft departing the PDM activity on one date must reenter that process at a fixed interval later.

Perhaps equally important, the aircraft within the PDM facility generally enter and leave the facility on a programmed schedule that is planned to minimize fluctuations in the number of aircraft in work—at least for a year or so. Thus, the PDM facility itself operates as if it were a closed facility with a fixed population of aircraft in work.

Such closed networks behave slightly differently from open networks, due to the fixed population within the network. Open queuing networks are often used to model service networks in which the number of customers varies considerably over a short time due to the unpredictability of arriving jobs. Thus, they are appropriate when modeling any kind of service process for which there is little or no control over job entry, such as banks or grocery stores.

In contrast, the operations and/or PDM process is a closed network. More important for our analysis, the PDM activity itself behaves as if it were closed, in the sense that jobs entering the activity are roughly synchronized with jobs departing.[1]

Queuing Versus "Ample Capacity" or Unrestricted Flow Models

Often, it is adequate for operations analysts to assume that the system under consideration can be characterized adequately by a network of delay times with unlimited capacity.[1] Typically, such systems are characterized by situations in which the "hands-on" process time is only a small fraction of the time required to complete the process. Those situations typically arise when there are ample resources to meet a wide range of workload arrival rates, or when there is a long delay that does not require significant resources. An example of the first is a sheet-metal shop whose workload is highly irregular: It may need to maintain a staff to meet its peak workload. An example of the latter is a paint shop, in which much of the flow time is spent waiting for the paint to dry.

The PDM activity requires large, expensive facilities and skilled technicians whose knowledge and expertise require years of practice and extensive training to develop. Most of the work is hands-on and therefore ties up a substantial amount of labor and the available facilities, with only a few tasks such as painting requiring less labor per unit of time. Because the capacity is difficult and expensive to acquire, it seems unlikely that there will ever be capacity ample enough to meet a sudden surge in workload. For this reason, we believe that ample capacity models cannot correctly estimate the effect of queuing (waiting for resources) on the number of aircraft in work if the workloads were to grow or if capacity were to change.

---

[1] Operations analysis is a process concerned with evaluating the financial, material, and capability effects of materiel or information flow processes. It uses a wide range of tools to estimate the quantities, location, and condition of material or information in large, complex systems, often as those values change dynamically.

Discrete Versus Analytic Queuing Models

Previous models have estimated work in process in two different ways: discrete simulation and analytic (mathematical) models. Discrete simulations use detailed state-transition rules for moving discrete jobs from one state (e.g., a workstation or dock) to another, process time probability distributions for the time a job stays in each state, detailed state recording, and statistical analyses of multiple simulated trials in a computer model of a process to conduct simulated experiments that allow one to compare the performance of a complex network of tasks and resources under alternative operating assumptions and processes. Examples of such systems include the Logistics Composite Model (LCOM) (Fisher et al., 1968), the Theater Simulation of Airbase Resources (TSAR) (Emerson, 1982), and the Dynamic Simulation of Intermediate Repair (DynaSIM) (Miller, Stanton, and Crawford, 1984). Computer simulation frameworks, computer languages, and graphical interfaces have been developed to support the definition of queuing simulations.[2]

Analytic models use detailed equations and mathematical transformations to represent the same queuing networks and use their detailed processes to estimate performance. To make these models tractable, their designers often make some simplifying assumptions about the probability distributions associated with the arrival and service process times, but they require essentially the same detailed information about the network flow probabilities as do the discrete simulations.

Both approaches have their place. Discrete simulations, such as LCOM, excel at measuring the effects of complicated resource or process interactions (such as minimal-size sorties or cross-utilization training rules), but they require a cut-and-try approach to identify an appropriate mix of resources. In contrast, analytic models operate much more rapidly and have the added advantage that the underlying mathematics can often be manipulated to identify an optimal mix of resources that minimize queuing delays or costs. If the focus is on how changes in detailed work rules might change production flows, discrete models are

---

[2] Many of these simulations present their results in some form of graphical display to facilitate quicker, more-accurate understandings of the patterns across different system designs.

more appropriate. If the focus is on identifying the most cost-effective balance of resources, analytic models can usually answer those questions more efficiently.

Limits of Both Approaches to PDM Modeling

Unfortunately, neither of these approaches is satisfactory for making long-term forecasts of PDM production and work in process—first, because of the large amount of detailed but unobtainable data required; second, because of the unforeseeable changes in those detailed data over time; and third, because of assumptions in the analytic models that traditionally limit their application to steady-state analyses or systems with "ample" capacity, i.e., those that will be unaffected by changes in flow time.

Both approaches require detailed state-transition probabilities, actual or assumed probability distributions for residence times at each server, and numbers of identical servers. Developing such a model requires a comprehensive, detailed collection of the work flows, resources, and work-scheduling rules—information that organic and contract PDM facilities regard as proprietary information that they would be reluctant to share with their competitors. As a result, it is difficult to obtain such information.

Even if it were possible to obtain such detailed information from observing current PDM operations, the mix of various tasks, the transition probabilities, and the time probability distributions are all subject to change over time. As new problems are found that require attention during the PDM process, those new tasks will add new work for particular job steps, while other tasks will disappear as material deficiencies are resolved. As these changes occur, the carefully crafted model of today's PDM processes will represent the future PDM process less and less accurately.

Finally, the traditional analytic models have restrictions that make them less generally applicable to the full range of queuing problems: They cannot accurately represent queuing under conditions in which the workloads change. More specifically, there are two general families of analytic models. The first family allows for dynamic changes in arrivals but assumes that those changes will have no effect on the overall flow times. (They are among the "ample" capacity models discussed previously.) The second family can estimate the queuing effects of different arrival rates for systems with fixed resources, but only if those rates remain constant over time.

The PDM process is one in which both capacity and dynamics are relevant to the estimation of flow times, production quantities, and aircraft in work. Thus, current models and standard modeling techniques do not appear to address systems, such as PDM, in which detailed process-flow data are unavailable and where there is at least some level of dynamic changes in the capacity or the arrival process.

BJB Analysis

Zahorjan et al. (1982) addressed the data-requirements problem when they discovered a way to estimate upper and lower bounds for analytical homogeneous and separable closed queuing networks,[3] based solely on the number of jobs in a system, the job stage loadings (time a job spends at each job stage, totaled across all visits to that stage), and the amount of time it would take one job to pass through the system if no other jobs were in progress. To use their method, one need not know the flow-time distributions, nor need one know the transition probabilities from one job stage to another.

Their approach applied queuing network models to computer system performance. The motivation for this approach is that in the early states of a computer model design, the expense of finding exact solutions is not warranted by the level of accuracy required.

---

[3] *Homogeneous networks* are defined as networks in which the workload requirements of every job entering a stage are equivalent. When applied to the multiserver case, this definition is expanded to include the servers, so that every job entering a stage has an equivalent workload that does not depend on the server.

Separable networks have an analytic solution, which may be found by using a product-form equation. A necessary and sufficient condition for the existence of product-form solutions is that the steady-state probabilities can be found by solving the local-balance equations. These equations balance the rate at which the continuous time Markov chain (CTMC) leaves a state with the rate at which the CTMC enters it. See Bolch et al. (1998, p. 283).

Until their work was done, the only bounding approximation was one that estimated an upper bound that used the smaller of two quantities:

$$X(N) = \min\left(\frac{N}{R_0}, \frac{1}{L_b}\right), \tag{A.1}$$

where

X = the production output rate
N = the number of jobs in the system
R_0 = the minimum (one job) flow time (i.e., no queuing)
L_b = the largest loading across all job stages.

Zahorjan and his colleagues called this previous approach *asymptotic bound analysis* (ABA), because the actual throughput approached the upper bound as the number of jobs in the system under study approached either one or infinity. It represents the two extremes of a queuing system, one with ample capacity (N/R_0), and one limited by the minimum time at the most highly loaded job stage, at which the capacity is overwhelmed. The only assumptions required for the ABA bounds to hold is that the device loadings are independent of the number of jobs in the system and that the job can occupy only one job stage at a time (sometimes called *separability*, whereby each node can be treated as a "separate" or independent queuing process).

They improved substantially on the ABA approach with a method they called *BJB analysis*, which produced both a tighter upper bound and a new lower bound. They begin by showing that the production capability of a balanced closed network (one in which all the loadings are equal) is exactly

$$X(N) = \frac{N}{(K + N - 1)L}, \tag{A.2}$$

where K is the number of job stages, and the other terms are as defined for Equation A.1.

This result holds true regardless of transition probabilities, network topology, or the number of visits required at each job stage. Recall that R_0 is the sum of the loadings, or the number of stages times the time spent in each. This calculation is more restrictive than the ABA calculation in that the result depends not only on job-stage separability but also on network homogeneity. Homogeneity requires that the routing between job stages be invariant (i.e., a job awaiting one job stage cannot jump ahead to another job stage), and that the job-stage service-time probability distributions be invariant regardless of the queue of jobs awaiting service.

They then obtained upper and lower bounds on throughput for a network with arbitrary loadings. The lower bound is trivially found by substituting the loading of the blocking stage (i.e., the largest loading) into Equation A.2. For the upper bound, they demonstrated that one can obtain an accurate upper bound by substituting the average loading across all devices for L in Equation A.2. Thus, they found that

$$\frac{N}{(N+K-1)\,L_b} \le X(N) \le \frac{N}{(N+K-1)\,L_a}, \qquad \text{(A.3)}$$

where

L_a = the average loading across all job stages
L_b = the maximum loading across all stages.

Of course, the previous bounds from the ABA model still apply. In the limit as N grows large, the output still cannot exceed $1/L_b$. A more complete expression that integrates both bounds would be

$$\frac{N}{(N+K-1)\,L_b} \le X(N) \le \min\left[\frac{N}{(N+K-1)L_a}, \frac{1}{L_b}\right]. \qquad \text{(A.4)}$$

But this equation requires routing homogeneity. As mentioned before, routing homogeneity requires that the job paths through the system be invariant with respect to the amounts of service time (loading) required at each server. In the PDM process, as in many other multistage flow processes, it is common to have several interchangeable servers at each job stage. In that circumstance, a job arriving at a particular stage can be routed to the first available server. To address this circumstance, RAND extended the Zahorjan et al. (1982) model to the more-general situation in which there are multiple servers at each job stage. As shown in Appendix B, RAND demonstrated that

$$\frac{N}{R_0 + L_b(N + C - K - 1)} < X(N) < \min\left[\frac{N}{R_0 + L_a(N-1)}, \frac{1}{L_b}\right], \tag{A.5}$$

where

C_k = the number of interchangeable servers (docks) at each stage k

$C = \sum_K c_k$, the total number of servers (docks)

$L'_k = \frac{L_k}{c_k}$, the average loading at stage k

$L'_b = \max\left(L'_k\right)$

$L_a = \frac{\sum_K L'_k}{C}$, the average L'_k across all stages k

R_0 = minimum hands-on flow time through all stages k.

This formulation relaxes one of the more constraining assumptions of the Zahorjan et al. BJB formulation: that of strict routing homogeneity. That is, when a job arrives at a particular job stage, it can be routed to the first available server, rather than waiting for a particular one.

One consequence of that relaxed assumption is that RAND's formulation is not exact, even for a perfectly balanced system. That is, the equation can only bound the throughput for a system, even one that is perfectly balanced (that is, one in which all the L'_k are identical, including the extreme case, in which all stages have an equal number of servers and an equal loading L_k). This is a direct consequence of the relaxed routing homogeneity assumption and the bounding assumptions about the probability of a job arriving at each stage encountering other waiting jobs.

Of course, RAND's bounds still require substantial information about each stage (the number of stages, the number of servers, and the loading at each stage). Again, we expect that this information will be difficult to obtain for the current system and unobtainable for the workload mix and PDM resource mix in the future.

To reduce the data requirements to the bare minimum, we observed that the PDM process managers operate in an environment in which cost and schedule are very important. To minimize their resource, flow time, and funding requirements, they strive continually to achieve a balanced system. An unbalanced system inevitably has idle resources waiting for jobs to arrive from the blocking job stage. By reallocating some of those idle resources to the blocking job stage, the managers can increase their resource utilization, reduce the requirement for additional resources and funds, and speed the flow through the blocking stage to improve throughput. While those reallocations are seldom perfect in practice, we assume that PDM managers will allocate their available resources to balance the system as tightly as possible.

With that assumption, every server throughout the system would have the same loading,

$$L_a = \frac{R_0}{C} \tag{A.6}$$

Note that, as described in Equation A.5, the equation may also be expressed as

$$L'_a = \frac{\sum_K L'_k}{C},$$

where

$$L'_k = \frac{L_k}{c_k}$$

$$C = \sum_K c_k .$$

This equation holds only if

$$\frac{L_1}{c_1} = \frac{L_2}{c_2} = \ldots = \frac{L_k}{c_k}.$$

This is true if managers act to balance workloads across all work centers so that each work center has approximately the same production rate.

Dynamic Workloads and Capacity

Finally, future workload forecasts may be constant or may ebb and flow over time. If the PDM arrival and service process were stable, one could use the Zahorjan-RAND formulation directly.

Unfortunately, the PDM process is replete with dynamic changes in workloads. Over the aircraft life cycle, the amount of PDM work per visit grows at an accelerating pace (Pyles, 2003). In addition, modifications are integrated with the PDM work from time to time. The production process, with different numbers of aircraft produced each year, causes the arrivals to cycle from year to year. Finally, the occasional change in the interval between PDM visits introduces temporary PDM-induction valleys that ripple through subsequent years.

To address more-dynamic problems, we numerically solve a time-based, nonlinear differential equation for which the Zahorjan-RAND BJB model is used to represent the production process. This approach requires two key assumptions: that the arrival process is continuous and that the changes over an integration interval are small enough that the number of aircraft before and after each time induction is "nearly" equal. The first assumption allows for changes in the arrival rate over time, but no sudden induction of many aircraft. This assumption approximates the PDM process quite well, wherein plans are developed to induct a number of aircraft over a year by spreading those inductions approximately equally over the year. Subject to limits in the numerical stability, one can make the integration interval arbitrarily small, thereby ensuring that the second assumption is valid.

Thus, we propose the following differential equation as a representation of the change in the number of aircraft in PDM at a particular time:

$$\frac{\partial N(t)}{\partial t} = \frac{-N(t)}{R_0(t) + \left[N(t) - 1\right] L_a(t)} + \frac{\partial P(t)}{\partial t}, \tag{A.7}$$

where $P(t)$ is the arrival rate during the integration interval and other terms are as previously defined. The first expression in the differential equation represents the PDM production rate at time t, while the second term is the PDM induction rate.[4]

The differential equation is nonlinear, and it does not have a closed-form solution. To solve it for practical problems, we chose to approximate it by performing a numerical integration in a spreadsheet model.[5] To make this computation tractable in a spreadsheet, we used a

---

[4] Note that we have focused here on the upper-bound result. Future work may address the results for the lower bound. Also, this expression has incorporated the balanced-system assumption from Equation A.6.

[5] Although this document does not discuss it, we have also solved this differential equation for a steady-state solution for the number of aircraft in PDM based on a given workload. That solution appears inferior dynamically because it overestimates the instantaneous effect

simple predictor-corrector integration method that computes the end-of-period (e.g., end of fiscal year) expected value of *N(t)* based on the period's expected initial value, the planned inductions, and an iteratively improved estimate of next year's value:

$$N_{t+1}^{(i)} = N_t + I_t - \frac{C_t}{2}\left\{\frac{N_t}{R_0^-\left(C_t + N_t - 1\right)} + \frac{N_{t+1}^{(i-1)}}{R_0^+\left[C_t + N_{t+1}^{(i-1)} - 1\right]}\right\} \quad \text{(A.8)}$$

where

$N_{t+1}^{(i)}$ = the i th iterative estimate of the PDM aircraft in work at time $t+1$

N_i = the final estimate of the PDM aircraft in work at time t

C_t = the capacity (e.g., number of PDM docks) available between t and $t+1$

R_0^- = the hands-on PDM time required for aircraft inducted between $t-1$ and t

R_0^+ = the hands-on PDM time required for aircraft inducted between t and $t+1$

I_t = the total number of aircraft to be inducted (smoothly) between t and $t+1$.

Note that R_0^- represents the hands-on time associated with the previous period's aircraft; R_0^+ represents the (possibly different) time for aircraft being inducted in the period being estimated

This method iteratively estimates sequentially more-accurate estimates of the number of aircraft in work at the end of a production

of a workload change. The differential equation yields more-believable (and more-moderate) transient behavior.

interval by using the sequentially more-accurate estimates in a trapezoidal approximation to the integration.

From this equation, one can also derive the number produced as

$$P_t = N_t + I_t - N_{t-1}, \qquad \text{(A.9)}$$

where P_t is the number produced (completed) between t and t+1. The number produced is then the number inducted at some later time,

$$I_{t+d} = P_t, \qquad \text{(A.10)}$$

where I_{t+d} is the number of aircraft inducted in some later time period, which is identical to the number produced (P_t) a delay time (d) earlier. (For example, a facility supporting an aircraft fleet on a five-year cycle would induct the same aircraft this year that were produced five years ago.) The practical result of this phenomenon is that any anomaly in outputs one year will have a ripple effect on subsequent years' inductions and completions.

Estimating Inductions When Induction Intervals Change

To enable evaluations of dynamic induction policies, the spreadsheet keeps an annual memory of the number of uninducted aircraft by period of last induction. Thus, it uses a first-come, first-served queue of historical annual production that is advanced one step each period. Each period's inductions are limited to those aircraft at or above the number of periods in a PDM interval.

Induction policies with noninteger values are allowed by inducting the fractional portion of each period. For example, if the induction interval policy were 5.25 years and the integration period was one year, the model would induct all aircraft that were last produced six (or more) years earlier and three-fourths of those produced five years earlier. If the interval were extended by a year, the model would not induct any new aircraft for the following year and inductions would then resume as before. If the period were shortened by a year, the oldest two years' previous production would be inducted in that year.

This mechanism is useful for examining the dynamic effects of changing the inspection interval. A controlled interval extension (CIE) that is too abrupt can create a series of peaks and valleys in subsequent years' workloads. While those cycles may diminish slowly over time, the instability can create substantial production management challenges.

Extending BJB Analysis to Multiple-Server Cases

A key relation used in the Zahorjan et al. (1982) model is the mean value analysis equation,

$$R_k(N) = [1 + \bar{n}(N-1)]L_k \ , \qquad \text{(B.1)}$$

to estimate the residence time, R_k, at node k. The expression indicates that the residence time is the sum of the service and queuing times of the Nth customer, the latter being given by the average queue length (with $N-1$ customers in the network) times the service time. The analysis assumes there is only one server at node k.[1]

For the more general case with c_k servers at node k, the time the Nth customer waits in the queue is more complicated. There are three cases: (1) if there are $n < c_k$ customers at node k, the arriving customer has no wait in the queue; (2) if there are $n = c_k$ customers at node k, the customer must wait in the queue for one customer to be served at rate c_k/L_k, where L_k is the loading at the kth node; (3) if there are $n > c_k$ customers at node k when the Nth customer arrives, they must wait in the queue until $n-c_k-1$ customers are serviced at rate c_k/L_k. This gives

[1] Another way to conceptualize this formulation is to imagine that you are the Nth customer entering a queue at the bank. The time you wait is equal to the time that the $N-1$ customers ahead of you must wait, plus your own service time. If there is only one server, the $N-1$ customers ahead of you must each wait L_k.

$$R_k(N) = [1 + \bar{n}(N-1) + \sum_{n=0}^{c_k-2} (c_k - 1 - n) p_k(n \mid N-1)] \frac{L_k}{c_k}, \qquad \text{(B.2)}$$

where $p_k(n \mid N-1)$ is the marginal probability that n customers are at node k when there are $N-1$ customers in the network (Gross, 1998). This probability can be calculated recursively, but that defeats the simplicity of the Zahorjan et al. (1982) model. There exist, however, bounds on $R_k(N)$. Since the third term in the brackets in Equation B.2 is nonnegative,

$$R_k(N) \geq [1 + \bar{n}(N-1)] \frac{L_k}{c_k}, \qquad \text{(B.3)}$$

using the constraint that

$$p_k \leq 1 \ \forall n$$

and that

$$\sum_{n=0}^{c_k-2} p_k(n \mid N-1) \leq 1, \qquad \text{(B.4)}$$

it can be shown that

$$\sum_{n=0}^{c_k-2} (c_k - 1 - n) p_k(n \mid N-1) \leq c_k - 1,$$

which gives the bound

$$R_k(N) < [1 + \bar{n}(N-1) + (c_k - 1)] \frac{L_k}{c_k}. \qquad \text{(B.5)}$$

Combining Equation B.5 with the analysis of Zahorjan et al. gives

$$\frac{N}{R_o + L_b(N-1) + L_b \sum_{k=1}^{K}(c_k - 1)} < X(N) < \frac{N}{R_o + L_a(N-1)} \tag{B.6}$$

where

$$L'_k = \frac{L_k}{c_k}$$

$$R_o = \sum_{k-1}^{K} c_k L'_k = \sum_{k-1}^{K} L_k = R_o$$

$$C = \sum_{k=1}^{K} c_k$$

$$L_b = \max\left(L'_k\right)$$

$$L_a = \frac{\sum_{k=1}^{K} L'_k}{C}$$

In effect, these bounds are quite extreme. The upper bound is the equivalent of a system in which the average job encounters no other jobs queued at any stage, so it waits only until the first of the several servers in that stage becomes available, or L_a. In contrast, the lower bound is the equivalent of a system in which the average job encounters several other jobs already waiting for service at each stage, so it must wait until the last server in the stage has completed the work in progress before it can start work. When those long queues are combined with the additional consideration that time between departures is L_b in every job stage, the total flow times can be quite long for an unbalanced system.

Finally, note that the last term in the denominator of the lower bound can be expressed more simply as $C{-}K$ and that the ABA upper bound still applies in cases in which the system is not perfectly balanced, giving a final result:

$$\frac{N}{R_o + L_b(N + C - K - 1)} < X(N) < \min\left[\frac{N}{R_o + L_a(N - 1)}, \frac{1}{L_b}\right]. \quad \text{(B.7)}$$

APPENDIX C

Estimating Parameters

This appendix is intended to provide at least one way to obtain the required data for PDMCAT. We demonstrate how to use historical production and work-in-process data to estimate not only the current R_0, but also to estimate how R_0 will change as workloads ebb and flow.

One of the things we discovered in this analysis was that modification workload changes had a different effect on R_0 than a similar-sized change in PDM workload, so we discuss how to estimate both effects and how to integrate them to estimate the parameters in PDMCAT.

Some Parameters Can Be Observed Directly

As one can see from Equation A.9 in Appendix A, one needs projected inductions (P_t), minimum hands-on flow times (R_0), capacity (C), the induction interval (I), and the initial number in work (N_t) for the projected time period. Not so clear in the equation, one must also choose an integration step size, which we selected as one month.[1]

The induction interval is spelled out in AFTO 00-25-4 (U.S. Air Force, 2003). While a few aircraft assigned to the corrosion-prone bases in the Pacific Air Forces have a four-year interval, the vast majority of the KC-135 aircraft have a five-year, or 60-month, interval, and our analysis assumed that all aircraft were on that schedule. As a matter of practice, the order allows for some modest deviation from those inter-

[1] This step size was satisfactory for this analysis. We have found that some workloads with greater dynamics (e.g., a new workload) often need a smaller step size.

vals, but the fleet SPD can provide information about planned changes, and AFMC's AFKS aircraft status file contains historical information about the entry and departure of aircraft from organic and contractor PDM facilities.

We obtained annual induction, production, and work-in-process values from the AFKS aircraft status file. That file contains codes indicating where each Air Force aircraft is located, what command possesses it, and the current activity to which it is assigned. We processed that file to obtain the annual PDM induction count, production count, and number of aircraft in depot maintenance or modification status at any of the four depot locations[2] during the period 1996 through 2003.[3]

Two other measures were derived from the other observations and these data. First, we estimated the capacity for each year between 1998 and 2002, based on published briefings from OK-ALC and the two contractor facilities. Then we used the capacity estimate and the AFKS data to estimate the minimum hands-on flow time. The following two subsections describe the particulars of those estimates.

Estimating Physical Facility Capacity

The PDM process has a clear physical capacity constraint: the number of docks at which aircraft can be disassembled, inspected, repaired, and tested. Equipment and other facilities are arranged to perform certain services at each dock, and only one aircraft can occupy a dock at a time. The docks themselves are grouped into job stages roughly like those shown in Figure 2.1, and there are a number of interchangeable docks within each job stage. While labor is also a potential production constraint, it is incorporated into the hands-on time estimates discussed below.

---

[2] KC-135 PDM has been carried out at a variety of locations: Tinker AFB, Oklahoma City, Oklahoma; Kelly AFB, San Antonio, Texas; McClellan AFB, Sacramento, California.; and PEMCO Aeroplex, Birmingham, Alabama.

[3] For the purposes of this analysis, we wished to exclude any depot possessions that were made for activities other than PDM. Visits to non-PDM facilities (e.g., for modifications) and depot visits shorter than 170 days (e.g., for UDLM) were excluded.

KC-135 PDMs currently occur at three locations: OK-ALC, BASC, and PEMCO (in Birmingham, Alabama). In 1998, PDM inductions at SM-ALC were terminated, and that work was redirected to the BASC facility. For this analysis, we modeled all facilities as though they were a single conglomerate facility.

Only limited data were available about the number of docks available for KC-135 PDMs. Based on briefing materials from OK-ALC, BASC, and PEMCO, we estimated that the three facilities had a combined physical capacity of 72 docks in 2003. (We found no relevant data about the prior operations at SM-ALC.) While the sites' production capacities undoubtedly vary over time, depending on the amount of work planned and the labor available, the physical capacity is relatively stable because adding capacity requires construction of additional covered workspace.[4]

Estimating Historical Minimum Hands-On Flow Time, R_0

We turn now to estimating the KC-135 PDM hands-on flow time. First, we discuss how to estimate the hands-on flow times from historical data. Then we develop a way to estimate how hands-on flow times may change in the future, based on PDM life-cycle workload forecasts and historical hands-on flow-time estimates.

It is easier to understand this effect when measuring the minimum hands-on flow time over the same period. Of course, it is not financially possible to measure hands-on time directly, because one would need to reduce the facility inductions to measure the hands-on time for several aircraft being worked on, one at a time. To estimate the minimum hands-on time, one can rearrange the terms of the produc-

---

[4] All three facilities had potential additional KC-135 PDM space currently occupied by other workloads. As long as those workloads continue, we anticipate that only 72 docks can be dedicated to KC-135 workloads. It is possible to model each depot's production separately and then add the results. We chose to model a consolidated system for two reasons. First, the work performed on each aircraft does not vary between depots and neither does the depot's work center configuration. But more important, methods used to record earned labor hours differ between the contractor-run facilities at BASC and PEMCO and those of the Air Force–managed facility at Oklahoma City.

tion difference equation (see Equation A.8) to solve for the hands-on time over some period:

$$R_0 = \frac{N_{t+1}\, C_t \left(C_t + N_{t+1} - 1\right)}{2\left\{P_t - \dfrac{C_t N_t}{2\left[R_0^{-}\left(C_t + N_t - 1\right)\right]}\right\}}, \tag{C.1}$$

where all terms are defined as before, except that P_t, N_t, and N_{t+1} are measured, not forecast, values.

Of course, this equation has a practical flaw: Succeeding estimates of R_0 depend on estimates of earlier periods' values. If a measurement error occurs in the R_0 value for one period, that error will propagate to affect estimates of all future values. (The size of the error may either grow or diminish in subsequent periods, depending on the relative values of P_t, N_t, and C_t.)

To avoid this potential measurement propagation error, we adopted a simpler, approximate estimate of R_0 that implicitly assumes that the hands-on time varies slowly over subsequent production periods:[5]

$$R_0 = \frac{C_t}{2P_t}\left(\frac{N_t}{C_t + N_t - 1} + \frac{N_{t+1}}{C_t + N_{t+1} - 1}\right). \tag{C.2}$$

That is, this equation assumes that aircraft work carried over between periods will experience the same hands-on time as the newly arriving aircraft. More precisely, it measures the average hands-on time for aircraft delivered during the period, not for aircraft inducted during the period. While this equation is not as appealing theoretically as Equation C.1 for a stable process in which only the workload per PDM changes over time, it does have the advantage of detecting and measuring any flow-time consequences of changes in the production process.

[5] Equation C.2 is an application of the predictor-corrector method; details are available in Royce and Hurt (1967).

Several Factors Contributed to a Change in Burn Rate and Flow Times

The new contractors' productivity probably improved steadily as they refined their maintenance process and the workforce familiarized itself with the KC-135's unique requirements. While BASC had been able to hire skilled personnel as the former San Antonio Air Logistics Center closed its C-5 PDM line, those personnel would have initially been unfamiliar with the details of the KC-135 PDM process. As with any other supplier starting a new production process, BASC undoubtedly encountered production delays and expenses that it was later able to reduce as its labor force gained experience with the new workload's idiosyncrasies. Perhaps as important, the managers and schedulers at the new facility would have been able to discover ways to smooth the production flows and improve the rate at which labor was applied. Such production "learning curve" efficiency growth is common across a wide range of industries. In this particular case, the growth was quite rapid once the first aircraft was completed. The new contractor needed almost two years to complete the first PDM, but it was able to deliver ten aircraft in the following 12 months. By the end of 2002, the two contractors had reduced the number of aircraft in work at their facilities to about 50 percent of the 1999 peak.

For its part, the organic facility at OK-ALC had already achieved many of the learning-curve efficiencies, having produced several thousand KC-135 PDMs in previous years. Nevertheless, it was able to reduce its work in process noticeably in FYs 2001 and 2002. Managers at that facility credit both process improvements and increased direct and indirect labor with achieving the improved flow times.

Finally, most KC-135 modification workloads were transferred to non-PDM contractor facilities for aircraft entering PDM in FY 2001 and later. Prior to that time, the KC-135 TSG had employed the traditional approach, integrating major modifications with maintenance tasks. The theory behind that practice was that the preparatory and reassembly work for the two tasks would be duplicated if they were performed separately, so integrating them eliminated the duplication of time, effort, and cost. OK-ALC managers reported that the modifications frequently interfered with other PDM work and often prevented or slowed work on multiple areas of the aircraft simultaneously. As a consequence, all but a few minor modification workloads were contracted out, beginning with aircraft inducted in FY 2001.

Together, these workload-reduction and efficiency-enhancement measures reduced the three facilities' combined hands-on flow times from 1996 through 2003, as shown in Figure C.1.

Integrating Labor-Application Rates and Changing Workloads to Estimate Future Minimum Hands-on Flow Times

All these changes affected either the amount of labor required or the facilities' ability to apply their available labor. That effect suggests that, if one could forecast the amount of PDM work required per aircraft

Figure C.1
KC-135 Hands-On Flow Time Estimates, 1996–2003

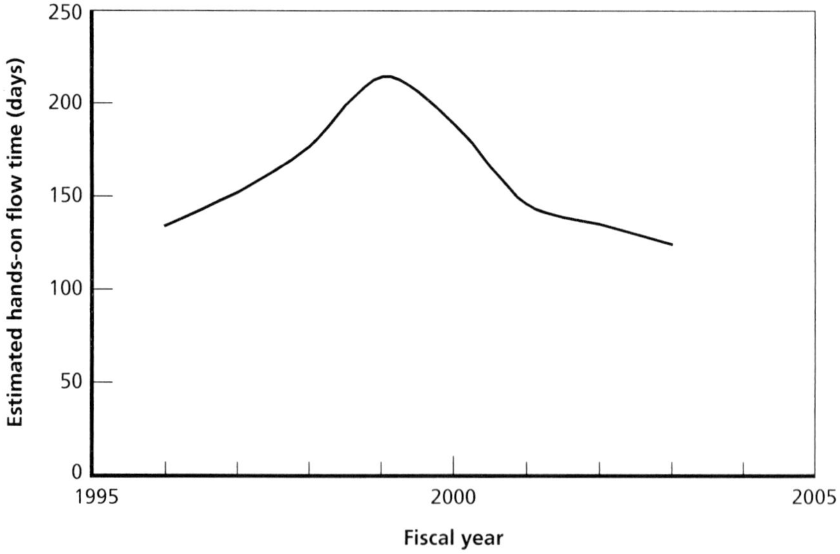

RAND *MG519-C.1*

and the rate at which labor could be applied, one could estimate the minimum hands-on flow time directly.

Estimating Future Workloads

Pyles (2003) found a general cross-fleet pattern for PDM growth as fleets age: The statistically derived PAF workload forecast is given by

$$\begin{aligned} W &= 13{,}300 - 111.3\,F + 301.3\,A - 424.2\,L \\ &\quad -1{,}287.8\,P + 8.572\,FA + 26.26\left(A-20\right)^2 \\ &\quad -7{,}868.8\,C_1 + 9{,}974.3\,C_2 - 2{,}578\,C_3, \end{aligned} \tag{C.3}$$

where

W = workload in depot product standard hours

F = flyaway cost, in millions of FY 1998 dollars

A = average age in years of the oldest mission design series

L = lag between the fleet of interest and the lead fleet

P = PDM interval in years

C_1 = categorical variable equal to 1 for a C-135 fleet but 0 otherwise

C_2 = categorical variable equal to 1 for a B-52 fleet but 0 otherwise

C_3 = categorical variable equal to 1 for a C-130 fleet but 0 otherwise.

In Equation C.3, the positive age terms imply that the amount of work will grow directly with the age of the aircraft. Indeed, the sixth term ($8.572\,FA$) implies that the more expensive aircraft will experience more-rapid workload growth, and the seventh term—$26.26\,(A–20)^2$—implies that the rate of growth will gradually accelerate throughout the aircraft's life cycle.

An analysis focused solely on the KC-135, the *KC-135 Economic Service Life Study* (Sperry et al., 2001), combined some regression analyses with interviews of engineering experts who also projected continued growth in KC-135 PDM workloads. Those experts identified several different workload types with different underlying growth rates, as shown in Table C.1.

The KC-135 office provided a more recent assessment of the PDM workload. The SPD's forecast of future PDM work closely resembles ESLS's, but with slightly fewer PDM hours. As shown in Figure 3.2, the ESL and SPD forecasts are very similar, but neither forecast agrees with the PAF forecast far into the future, in part because there is uncertainty about how the PDM workloads will grow when fleets are kept past the age at which the Air Force has prior experience with them.[6] The Air Force simply has not yet operated aircraft much past 45 years, but current plans call for operating portions of the KC-135 fleet until about age 80.

Table C.1
KC-135 ESLS Workload and Growth Forecasts

Workload Category	Initial Workload (DPSH)	Most Probable Case (percent)
Basic	14,540	2.4 until 2004[a] 1.0 thereafter[a]
MSR	2,000[b]	MSRs: 0.7/40 per year DPSH/MSR: 1 per year[a]
Heavy structure	6,400	2.4[a]
Over and above	677	5.0[c]
Concurrent modifications	435	2.4[c]
Rewire	5,000	—[d]
One-time tasks	2,500	0.0[e]
Economy	2,000	0.0[e]

[a] Linear.
[b] Two MSRs at 1,100 DPSH each.
[c] Compound.
[d] Complete in FY 2003.
[e] No growth.

[6] The sudden FY 2003 reduction in the ESLS workload is due to the forecast reduction in the rewiring workload (Sperry et al., 2001). Engineering development is under way for the third and final phase of this major maintenance program (Montgomery, 2003).

References

Bergren, Scott C., statement before the House Armed Services Committee, 107th Cong., 1st. Sess., Washington, D.C., March 23, 2001.

Bolch, Gunter, S. Greiner, H. de Meer, and K. S. Trivedi, *Queueing Networks and Markov Chains*, New York: John Wiley and Sons, Inc., 1998.

Chandler, Jerome Greer, "Putting PEMCO Back Together," *Overhaul and Maintenance*, March 2001.

Dietz, Dennis C., and Richard C. Jenkins, "Analysis of Aircraft Sortie Generation with the Use of a Fork-Join Queueing Network Model," *Naval Research Logistics*, Vol. 44, 1997, pp. 153–164.

Emerson, Donald E., *An Introduction to the TSAR Simulation Program: Model Features and Logic*, Santa Monica, Calif.: RAND Corporation, R-2584-AF, 1982. As of January 16, 2007:
http://www.rand.org/pubs/reports/R2584/

Fisher, Rolland R., W. W. Drake, J. J. Delfausse, A. J. Clark, and Al Buchanan, *The Logistics Composite Model: An Overall View*, Santa Monica, Calif.: RAND Corporation, RM-5544-PR, 1968.

GAO—*See* U.S. Government Accountability Office.

Gross, Donald, and Carl M. Harris, *Fundamentals of Queueing Theory*, 2nd ed., New York: John Wiley and Sons, Inc., 1985.

Jackson, James R., "Networks of Waiting Lines," *Operations Research*, Vol.5, No. 4, August 1957, pp. 518–521.

Johnson, Maj Gen Charles, "KC-135 Depot Status Reduction Plan," briefing, Oklahoma City Air Logistics Center, Tinker Air Force Base, Okla., December 2000.

KC-135 Tanker Sustainment Support Group, "Tanker Workload and Cost Forecast Workbook," spreadsheet, Tinker Air Force Base, Okla.: Oklahoma City Air Logistics Center, March 14, 2005.

Keating, Edward G., and Matthew C. Dixon, *Investigating Optimal Replacement of Aging Air Force Systems*, Santa Monica, Calif.: RAND Corporation, MR-1763-AF, 2003. As of January 16, 2007:
http://www.rand.org/pubs/monograph_reports/MR1763/

Keating, Edward G., Don Snyder, and Elvira N. Loredo, *Aging Aircraft Repair-Replacement Decisions with Depot-Level Capacity as a Policy Choice Variable*, Santa Monica, Calif.: RAND Corporation, MG-241-AF, 2004. As of January 16, 2007:
http://www.rand.org/pubs/monographs/MG241/

Kennedy, Michael, Laura H. Baldwin, Michael Boito, Katherine M. Calef, James S. Chow, Joan Cornuet, Mel Eisman, Chris Fitzmartin, Jean R. Gebman, Elham Ghashghai, Jeff Hagen, Thomas Hamilton, Gregory G. Hildebrandt, Yool Kim, Robert S. Leonard, Rosalind Lewis, Elvira Loredo, Daniel M. Norton, David T. Orletsky, Harold Scott Perdue, Raymond A. Pyles, Timothy L. Ramey, Charles Robert Roll, Jr., William Stanley, John Stillion, Fred Timson, and John Tonkinson, *Analysis of Alternatives (AoA) for KC-135 Recapitalization: Executive Summary*, Santa Monica, Calif.: RAND Corporation, MG-495-AF, 2006. As of January 16, 2007:
http://www.rand.org/pubs/monographs/MG495/

Miller, Louis W., Richard E. Stanton, and Gordon Crawford, *Dyna-SIM: A Nonstationary Queuing Simulation with Application to the Automated Test Equipment Problem,* Santa Monica, Calif.: RAND Corporation, N-2087-AF, 1984. As of January 16, 2007:
http://www.rand.org/pubs/notes/N2087/

Montgomery, Gerald, "Other High-Interest Programs," briefing, KC-135 System Program Directorate, Oklahoma Air Logistics Center, Tinker Air Force Base, Okla., February 12, 2003.

Pyles, Raymond A., *Aging Aircraft USAF Workload and Material Consumption Life Cycle Patterns*, Santa Monica, Calif.: RAND Corporation, MR-1641-AF, 2003. As of January 16, 2007:
http://www.rand.org/pubs/monograph_reports/MR1641/

Royce, Beckett, and James Hurt, *Numerical Calculations and Algorithms*, New York: McGraw-Hill, 1967.

Sperry, Kenneth, and the KC-135 ESL Integrated Product Team, *KC-135 Economic Service Life Study*, Tinker Air Force Base, Okla., EA 00-023R2135OTH, 2001.

U.S. Air Force, "Depot Maintenance of Aerospace Vehicles and Training Equipment," Change 1, Washington, D.C., Air Force Technical Order 00-25-4, April 15, 2003.

U.S. Government Accountability Office [formerly General Accounting Office], *Air Force Depot Maintenance: Improved Pricing and Cost Reduction Practices Needed*, Washington, D.C., GAO-04-498, June 2004.

Zahorjan, John, Kenneth C. Sevcik, Derek L. Eager, and Bruce Galler, "Balanced Job Bound Analysis of Queuing Networks," *Communications of the ACM*, Vol. 25, No. 2, February 1982, pp. 134–141.